10 0708650 7
UNIVERSITY OF NOTTINGHAM
WITHDRAWN
FROM THE LIBRARY

What Editors Want

D1350091

On Writing, Editing, and Publishing
Jacques Barzun

Telling About Society
Howard S. Becker

Tricks of the Trade
Howard S. Becker

Writing for Social Scientists
Howard S. Becker

Permissions, A Survival Guide
Susan M. Bielstein

The Craft of Translation
John Biguenet and Rainer Schulte, editors

The Craft of Research
Wayne C. Booth, Gregory G. Colomb, and Joseph M. Williams

The Dramatic Writer's Companion
Will Dunne

Glossary of Typesetting Terms
Richard Eckersley, Richard Angstadt, Charles M. Ellerston, Richard Hendel, Naomi B. Pascal, and Anita Walker Scott

Writing Ethnographic Fieldnotes
Robert M. Emerson, Rachel I. Fretz, and Linda L. Shaw

Legal Writing in Plain English
Bryan A. Garner

From Dissertation to Book
William Germano

Getting It Published
William Germano

The Craft of Scientific Communication
Joseph E. Harmon and Alan G. Gross

Storycraft
Jack Hart

A Poet's Guide to Poetry
Mary Kinzie

The Chicago Guide to Collaborative Ethnography
Luke Eric Lassiter

How to Write a BA Thesis
Charles Lipson

Cite Right
Charles Lipson

The Chicago Guide to Writing about Multivariate Analysis
Jane E. Miller

The Chicago Guide to Writing about Numbers
Jane E. Miller

Mapping It Out
Mark Monmonier

The Chicago Guide to Communicating Science
Scott L. Montgomery

Indexing Books
Nancy C. Mulvany

Developmental Editing
Scott Norton

Getting into Print
Walter W. Powell

The Subversive Copy Editor
Carol Fisher Saller

A Manual for Writers of Research Papers, Theses, and Dissertations
Kate L. Turabian

Student's Guide for Writing College Papers
Kate L. Turabian

Tales of the Field
John Van Maanen

Style
Joseph M. Williams

A Handbook of Biological Illustration
Frances W. Zweifel

PHILIPPA J. BENSON
& SUSAN C. SILVER

What Editors Want

An Author's Guide to Scientific Journal Publishing

UNIVERSITY OF NOTTINGHAM
WITHDRAWN
FROM THE LIBRARY
GEORGE GREEN LIBRARY OF
SCIENCE AND ENGINEERING

The University of Chicago Press Chicago and London

Philippa J. Benson, Ph.D., is director of education and author services for The Charlesworth Group, an international organization that supports publishers. She has taught science writing and technical communication at Carnegie Mellon University, Georgetown, the United Nations, and the National Institutes of Health, and has launched scientific publications for two conservation organizations.

Susan C. Silver, Ph.D., is editor-in-chief of *Frontiers in Ecology and the Environment*, published by the Ecological Society of America. Previously, she held editorial positions at Academic Press and the British Dental Association and was editor of *Biologist* and *The Lancet Oncology*.

The University of Chicago Press, Chicago 60637
The University of Chicago Press, Ltd., London
© 2013 by The University of Chicago
All rights reserved. Published 2013.
Printed in the United States of America

22 21 20 19 18 17 16 15 14 13   1 2 3 4 5

ISBN-13: 978-0-226-04313-5 (cloth)
ISBN-13: 978-0-226-04314-2 (paper)
ISBN-13: 978-0-226-04315-9 (e-book)
ISBN-10: 0-226-04313-4 (cloth)
ISBN-10: 0-226-04314-2 (paper)
ISBN-10: 0-226-04315-0 (e-book)
1007086507
Library of Congress Cataloging-in-Publication Data

Benson, Philippa Jane, author.
What editors want : an author's guide to scientific journal publishing / Philippa J. Benson & Susan C. Silver.
pages cm. — (On writing, editing, and publishing)
Includes bibliographical references and index.
ISBN-13: 978-0-226-04313-5 (cloth : alkaline paper)
ISBN-10: 0-226-04313-4 (cloth : alkaline paper)
ISBN-13: 978-0-226-04314-2 (paperback : alkaline paper)
ISBN-10: 0-226-04314-2 (paperback : alkaline paper)
[etc.]
1. Scholarly publishing—Handbooks, manuals, etc. 2. Science publishing—Handbooks, manuals, etc. 3. Authors and publishers. I. Silver, Susan C., author. II. Title. III. Series: Chicago guides to writing, editing, and publishing.
Z286.S37B467 2013
070.5—dc23

2012010479

♾ This paper meets the requirements of ANSI/NISO Z39.48-1992 (Permanence of Paper).

This book is dedicated to our husbands Benjamin Xu and David Currie, for all their support and patience along the way.

Contents

Acknowledgments

We wish to thank all our Chinese hosts, who invited us to give the workshops in mainland China on which this book is based. Lindsay Haddon (British Ecological Society) and Laura Meyerson (Rhode Island University) were additional lecturers for the workshops and generously allowed us to incorporate some of their material.

We are also grateful to all the authors of the sidebar pieces, which we believe greatly enhance the book. We are similarly grateful to the Editors of the journals that provided data for figure 2.1 and to Bernie Taylor, who designed and produced the illustrations.

Finally, we are grateful to early readers Alex Soars (Brookhaven National Lab), Marco Sandrini (National Institutes of Health), Susan Ambrose (Carnegie Mellon University), Marie McVeigh (Thomson Reuters), Margaret Reich (Reich Consulting), and Chang Jie (Zhejiang University), as well as to our Editor at the University of Chicago Press, David Morrow, and to Jennifer Kuhn for administrative assistance.

CHAPTER ONE

Who cares what Editors want?

Some researchers believe that becoming expert in their science is the only important aspect of their professional development. The reality is that to become a world-class scientist today one must also be able to navigate the publishing process with skill and speed, as well as write with clarity, accuracy, and grace.

MONICA BRADFORD, Executive Editor, *Science*

Researchers in scientific, technical, and medical (STM) fields around the world study a diverse array of topics, but they all focus on the same professional goal: getting their science published. Despite efforts in academic circles and elsewhere to develop a broad spectrum of measures to evaluate a researcher's worthiness for employment or promotion, the age-old dictum still holds true: it's publish or perish.

Inexperienced authors often need help as they try to tackle the different phases of the publishing process—and sometimes during the earlier stages of manuscript preparation as well. They need guidance on how to judge when to write up their research and what kind of scientific papers to write. They need to learn how to select the journal that is most likely to publish a specific paper and how to submit their work to particular publications. When they receive a response to their submission, whether positive or not, they need advice on how to reply and what steps to take next. Some inexperienced authors turn to research advisors, colleagues, or academic mentors to get this advice. However, some advisors tend to focus more on helping students do the science rather than on how to write it up and submit it for publication. In reality, some advisors have received little advice and guidance themselves, have learned the publishing process by trial and error, and therefore aren't entirely comfortable leading others down that road. Sadly, a few senior researchers are more concerned with their own publishing projects than with those of their students and are unwilling to spend much time on mentoring in this area. Colleagues within the same field are also potential rivals and may be hesitant to share the tips and tricks they've learned.

In short, the imperative to publish never stops, but only fades slightly with time and tenure. The bottom line is that little formal training is available for researchers in this vital aspect of career development. As a result, many young researchers have almost no idea of what obstacles they'll encounter, and how to get to the finish line—a published paper—as quickly and painlessly as possible.

The aim of this book is to address the needs of these novice authors, whether graduate students, postdocs, young researchers, or faculty, to help them meet the specific challenges they may encounter at each stage of the publishing process. We also hope this book will be useful as a reference for senior researchers, as well as for teachers of science writing and for those who train up-and-coming Editors. However, we must stress here that Editors, and their opinions on how things should be done, are as varied as the journals they work on, a reality that became very clear to us as we heard from editorial colleagues while writing this book. For example, some Editors firmly insisted that authors should include line numbers when submitting a manuscript, as this saves reviewers a lot of trouble when drafting their reports, while others felt that line numbers are unnecessary. In yet other cases, authors are instructed not to add line numbers because the manuscript submission systems of some journals add these automatically. Similarly, one Editor claimed never to read cover letters, while another believed that it was very important for Editors to do so. (Our advice on these two issues is to always include line numbers on your manuscript *unless instructed not to do so in the instructions to authors of the particular journal you are submitting to* and to always provide a cover letter.) The point is that the opinions and preferences of Editors vary widely on many publishing matters.

CHANGING PERSPECTIVES

To help young authors understand the publication process, we first explain what Editors are looking for when considering new submissions and then examine the publication process in a logical sequence to reveal the reasoning behind the many requirements that journal Editors have. Few resources are available that describe what goes on in editorial offices, from who looks at initial submissions, to what criteria are used to separate submissions rejected without review from those that are sent to peer review, to how the peer review process itself works.

Novice authors are often not aware of how different behind-the-scenes operations can be from journal to journal. No two journals are exactly alike in terms of the criteria used to select papers or the processes used to assign reviewers. Editorial standards and styles also vary widely, as do the roles and responsibilities of Editors, other editorial office staff, and reviewers. Many of the top-tier journals (e.g., *Nature*, *Science*, *Cell*) and their associated families of specialty titles have full-time, professional Editors in their offices, working on their journals and making decisions, while the vast majority of academic journals have an off-site, academic Editor, working in a university or research institute, and doing journal work in his or her spare time. Operations also differ depending on the size, finances, and goals of each journal, and whether it is published by a not-for-profit professional society or a for-profit publishing house. In short, there is no exact formula for predicting what the Editor of a particular journal wants. However, there are ways of figuring this out if you know what to look for and where to look for it.

WHERE WE BEGIN

To provide a broad view of the editorial landscape of scientific publishing, we begin with a few assumptions. We assume, for example, that researchers have completed their research and have correctly analyzed their findings. We also assume that the research is original, that the findings are valid, and that the science has been done properly, adhering to all legal and ethical guidelines, although we do touch briefly on scientific misconduct and ethical issues (see chapter 11). We also assume that some readers will be nonnative speakers of English, since, in recent years, journals have seen huge increases in submissions from international authors, with the greatest numbers coming from countries such as China, Korea, Japan, India, and Brazil (figure 1.1).

We also acknowledge that there may be limits to how widely readers can apply the perspectives and information we provide, since we are drawing on our own experiences working in editorial offices in the US and the UK. In other parts of the world, editorial practices may be different and researchers living in non-Western cultures may be accustomed to editorial requirements that differ from those associated with Western English-language journals.

We have also limited the scope of this book to the processes of publishing a manuscript and therefore do not provide specific instructions on how to write a scientific paper. Many excellent textbooks on technical and science writing are available, and many universities and colleges also provide free online resources that offer guidelines and examples of how to write clear and readable technical prose. We do touch on the importance of authors knowing when they are ready to start drafting a manuscript, but this is in the context of better understanding what Editors want. In appendix 1, we

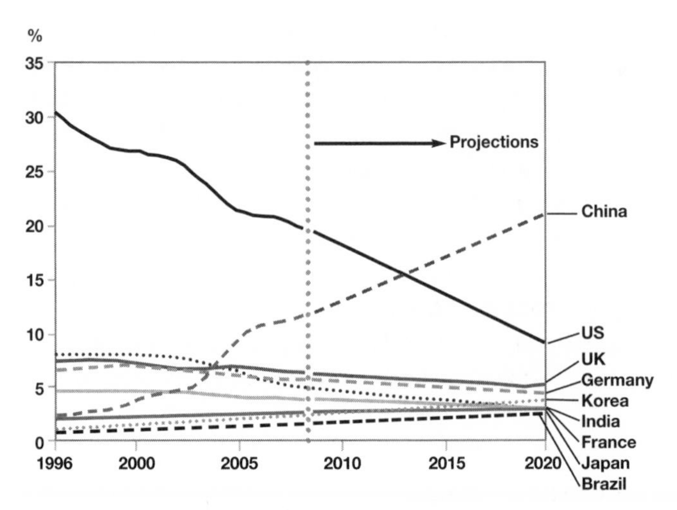

FIGURE 1.1 Linear extrapolation of future publishing trends. With permission from the Royal Society. Source: The Royal Society, *Knowledge, Networks and Nations*, 2011.

include a list of some of the many resources that are available, including links to online writing resources offered by universities and colleges across the US.

In this book, we also occasionally note information that is specifically for nonnative speakers of English. While writing a paper in a foreign language is an added challenge for authors, we believe that many of the other challenges in publishing are actually the same for both native and nonnative speakers. Certainly, if a manuscript is written in clear, concise, and readable prose, Editors will have an easier time assessing the quality and appropriateness of the paper for their journal. However, success in publishing comes not only from good writing but also from making sound choices about when, where, and how to submit a paper for consideration.

Throughout the text, we have also included sidebars by a variety of experts, including Editors and publishers from well-known journals and publishing houses. These contributors provide their views on a broad range of topics that are central to scientific publishing and add a variety of explanations and opinions that are both interesting and informative.

WHAT WE COVER

We begin chapter 2 by explaining how an understanding of an Editor's perspective can help authors more successfully navigate the publishing process. In chapter 3, we discuss how authors can judge whether they are actually ready to write a full scientific paper or whether they should consider waiting or using a different channel of communication at a particular stage of their research. The decision regarding when and where to submit a paper for publication is where our story really begins because how that decision is made is crucial. We move on in chapter 4 to discuss issues of authorship and the importance of agreeing on roles, responsibilities, and author order at a very early stage of manuscript preparation.

In chapter 5, we discuss how authors can narrow down the choice of journals to two or three possibilities and stress the importance of thoroughly studying previous issues of each of the short-listed publications, as well as their websites and instructions to authors. Together, these resources should provide not only the form and format that Editors want but also the scope of the journal, the topics it will consider, and the specific approaches to the science that the Editors are looking for. Although a complete and clear set of instructions to authors is perhaps the best resource that authors have, in reality these instructions are not always as coherent and comprehensive as authors might wish. In fact, in the past few years, science Editors are increasingly addressing the challenge of improving instructions to authors through their professional organizations, such as the Council of Science Editors (CSE, see www.councilscienceeditors.org), the Society for Scholarly Publishing (SSP, see www.sspnet.org), and the International Society of Managing and Technical Editors (ISMTE, see www.ismte.org). Chapter 6 discusses how impact factors are calculated and what they really mean because so many authors are influenced by the impact factor of a publication. In chapter 7 we give advice on how to write a cover letter and in chapter 8 we provide guidelines on how to prepare a manuscript for submission to a journal and how to avoid mistakes that authors often make when submitting papers.

Chapter 9 reviews general information about how peer review is managed, including brief descriptions of what different editorial staff do and who authors should approach with particular questions or problems. We then explain the range of editorial responses an author may receive in decision letters from journals, what those responses mean, and how to respond to them.

In the two final chapters, we look at some of the ethical problems that authors and Editors have to grapple with, including plagiarism, copyright issues, and multiple submissions. We conclude with a discussion of various trends in the publishing industry and how they may affect authors—these include open access, new measures of impact, and the way new technologies might change the way publications are produced and read.

The appendices at the end of the book provide readers with additional resources that we hope will help them succeed in their publishing efforts.

THE BOTTOM LINE

We titled this first chapter "Who cares what Editors want?" and we hope that by the time you finish this book you will understand why it is that you, the author, need to care what Editors want. We also hope that the explanations, tips, tools, references, and resources we provide will help you to better understand the roles and goals of journal Editors and, in turn, how to publish your research with more confidence and success. That's what you want, and that's what Editors want too.

CHAPTER TWO

Changing perspective from author to Editor

Those who can tell us the most about journal publishing are the editors, whose success as authors and/or reviewers secured their appointments as editors. With their unique perspective, no other group is better prepared to advise on how to effectively play the publishing game. KNAPP AND DALY, *A Guide to Publishing in Scholarly Communication Journals*

Competition to publish is stiff, and scientific careers depend on a researcher's success in publishing papers in well-respected scientific journals. Along with conducting and publishing studies, researchers also have to keep up with new techniques and technologies, go to meetings (often in faraway places), and learn to use new tools for communicating and collaborating with colleagues. Many researchers are faculty members and so must also develop and teach multiple courses to undergraduate students and perhaps supervise graduate students and postdocs as well. It's no wonder that most researchers are eager for a set of clear guidelines to help them succeed in publishing their science.

Many scientists become deeply engaged in the complexities of their work and believe that other people are (or should be) equally intrigued by the problems they are trying to solve. This concentrated focus on a narrow area of research can lead to problems in publishing because when researchers write, they often assume that readers share their deep knowledge of and interest in the topic of the paper. Unfortunately, neither the researchers' fascination with their work nor their desire for a clear-cut recipe for success in publishing is of much help in actually getting published. Authors face an exploding number of channels for finding and reading information—from print and online publications to blogs and twitter feeds—and an almost equal number of new tools and technologies for writing and publishing. The challenge for authors is to learn how to navigate their options and target the best medium for presenting their new findings. When the chosen medium is a scientific journal, we believe that if authors can learn to see the publication process from the perspective of a journal Editor, they will be more successful in getting papers into and through peer review to publication.

The difference between the perspective of an author and that of an Editor can be thought of in much the same way as the difference between a programmer and a user of software. The programmer who writes the software understands exactly how the code was written, why it was written that way, and how the program is supposed to work. Users don't usually care about why a program was written in a particular way or how the code works; they care only that the program helps them achieve their goals and that it makes their work easier along the way. Similarly, authors may do their science meticulously and believe strongly in its importance, but the Editors of the journals they submit to don't necessarily share the same background knowledge or the commitment to any particular aspect of research. Journal Editors have their own jobs to do and have very specific hopes for each new submission that arrives: "Will this paper interest our readers and advance our knowledge of the field? Will it improve our coverage of this particular topic in the journal? Will it increase the reputation of my journal and help to improve its impact factor?" Authors need to see Editors as their "users," making sure that the paper matches this particular user's requirements and that it is clear from the moment a file is opened that the paper is a good fit for the publication.

THE AUTHOR'S POINT OF VIEW

Many inexperienced authors see the publication process as beginning at the point when they submit a paper to a journal. However, decisions about where and when to publish should start as soon as you've completed your research, if not before. Key steps to success involve being realistic in recognizing what you've got in terms of new scientific knowledge, identifying the audience who will want to know about your work, and finding out which journals are read by that particular audience. Many authors write with the assumption that most (or at least many) of their end readers will be familiar with the concepts and terminology associated with their research, and so write for researchers who have some level of expertise in their field. While this may be true of readers of some narrowly focused journals, it is not so for readers of journals that cover a wider range of topics. Authors often fail to understand that the Editor is their first reader—the gatekeeper for all the other readers of that publication. Editors know their readers well and understand what they expect to find in the journal. The Editor is the first filter, the person who is responsible for picking the best offerings and discarding the rest.

Authors need to understand that it is the journal Editor who matters, as this is *the* person who will decide whether the paper should or should not go forward to peer review. Editors cannot be experts in every area that their journal covers and may not recognize the importance of every paper that comes in. The author's job is to intrigue the Editor, and later on the reviewers, and convince them of the relevance of their work.

THE EDITOR'S POINT OF VIEW

Most authors know that Editors of the most prominent high-impact-factor journals reject a large percentage of papers without review. Even when submitting to middle-ranked journals with only moderate impact factors, authors know that rejection rates can still be quite high. But the question is why: what are Editors thinking about when they begin to evaluate a newly submitted paper for possible peer review?

To begin seeing things from the Editor's point of view, we start with the assumption that the Editor is working in a relatively modern editorial office, using standard technologies, including an up-to-date online manuscript tracking and peer review system. Generally, the person who first processes incoming manuscripts (often one of several different types of Editor) sees a paper as an upload to the journal's online submission system, which has been assigned a permanent manuscript number. The Editor with the job of making a first "cut" will have a number of possible routes they can take in deciding whether or not to send the paper to peer review. Some Editors look at the cover letter first (if there is one), while others will go straight to the title and abstract. From there, some Editors will look at the results section, while others will jump straight to the discussion or conclusions.

Even though the approach may differ, Editors' initial questions remain the same: is this paper a good match for the journal? Does it fit the specified criteria, as spelled out in the instructions to authors? Is the science novel, relevant, and timely? Is the writing clear and concise? Some Editors may also consider whether the science goes beyond being original to ask if it is actually at the cutting edge of the field and will contribute to a current hot debate. Fraud in scientific studies is being uncovered more frequently these days and so Editors are also scrutinizing data and figures more carefully in initial review, to make sure experiments seem properly constructed and the resulting data look accurate and plausible. Editors may also take into account what they know about the authors and whether they have a good publishing record.

All of these are common questions that might go through Editors' minds

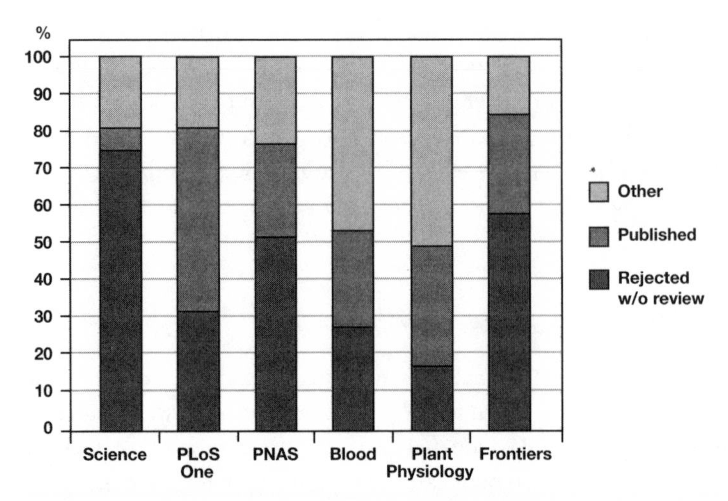

FIGURE 2.1 Data on manuscripts from six high-impact-factor journals in 2010.

when they are first deciding whether a paper should or should not go into the reject-without-review pile. Depending on the size of the journal, Editors may have to make this initial judgment for between ten to one hundred manuscripts each week and so must develop a strong and clear idea of what they are and are not looking for in a manuscript.

Surprisingly, many initial submissions don't meet these basic criteria. Often papers have notable problems: either they are outside the scope of the journal, or they reflect science that is neither novel nor interesting, or they have been sent in the wrong format or are missing important information (see chapter 7). These submissions go straight onto the reject-without-review pile or are sent back to the author for revision (figure 2.1).

What happens next to papers that do make it through the first cut depends on the editorial office structure and the size of the journal (see chapter 9). For small journals, the Editor may send a manuscript that has made the first cut directly to peer reviewers. For larger journals, Editors may instead send it to an Associate Editor (sometimes referred to as a section Editor or subject matter Editor), who then selects appropriate peer reviewers and sends the paper on to them. Every journal is different, but one way or another, once a paper has made it past the reject pile, the Editor will be faced with a new set of challenges and decisions, from finding suitable peer reviewers to assessing multiple submissions on a similar topic. Some papers fly through peer review while others move at a snail's pace, for a variety of reasons, ranging from the size of the journal to the availability of appropriate reviewers, even to the time of year. At the same time that the reviews are going on, Editors are juggling a multitude of other tasks: ensuring that enough other papers are moving through peer review and are being readied for publication, keeping issues coming out on schedule, assessing and maintaining high publishing standards, setting and monitoring budgets, dealing with editorial boards, managing staff, and meeting a host of other professional responsibilities.

THE BOTTOM LINE

As an author, anything you can do to make the Editor's job easier is going to be a positive step in moving the Editor towards deciding that your paper should be sent out to peer review. Anything that makes your manuscript problematic for the Editor—from wrong formatting to bad writing to unlikely data—will negatively influence the Editor's decision about whether to send your paper out for peer review. You will assuredly benefit by keeping the Editor's needs in mind as you embark on the road to getting your paper published.

CHAPTER THREE

Judging the newness of your science

As with computing, so with publishing: the greater the transparency, the better the scholarship. Scholars should open up as much as they can: data, tools, processes, results, and even paths not taken. Don't be scared of being wrong: be scared of obscuring your message. And be terrified of saying nothing.

JON ORWANT, Engineering Manager, Google

More than anything else, Editors want the manuscripts they receive to contain excellent, innovative science. This focus on publishing research that contributes to our understanding of a particular subject is not entirely altruistic; when a journal publishes a paper with research that will influence the thinking of other scientists, that paper is likely to get cited frequently, leading to a higher impact factor and a better reputation for the journal. All Editors wants their journals to be influential and well-respected.

TO PUBLISH OR NOT TO PUBLISH?

Throughout the ages, philosophers and teachers of science have recognized the centrality of communication in the scientific process and the fact that sharing ideas is fundamental to the advancement of knowledge in every field. "Publish or perish" is the mantra of scientific success because the process of peer review and eventual publication, in theory, leads to the weeding out of poorly conceived scientific theories and experiments, separating those that can withstand detailed scrutiny from those that cannot.

The need to publish scientific papers puts tremendous pressure on authors to get their work into print quickly and often. Those who publish first get most of the credit for new information or ideas, even if they were not the first to conceptualize or prove a new theory. Researchers also need to publish papers to be viable on the job market, to compete for funding, and to maintain their position at the cutting edge in rapidly developing fields. Obviously, academic departments and research institutions also look for researchers with extensive publication lists, since they are also in competition for funding and want to attract the most talented researchers on the market.

Sometimes, the pressure to publish is a positive force, keeping researchers focused on a clear goal. At the same time, the extreme pressure to publish in today's scientific arena can have very negative results, including unethical behaviors ranging from plagiarism to falsifying data (see chapter 11), But such behaviors are the extremes. A more common, but ill-advised response is to try to publish research before it has advanced far enough to be published.

Young researchers do not always have a clear view of when their work is developed enough to be ready for publication and rely on the advice of supervisors or senior colleagues. Often, knowing when research is ready to be written up is based on experience, or, in the case of young researchers, on the advice of supervisors or senior colleagues. Recognizing when a paper is ready for publication requires becoming familiar with all the possible venues for communicating new science and being able to select the one that will best carry the information to appropriate audiences. Authors also require a thorough knowledge of current understanding of a particular topic and a realistic appraisal of the potential of their work to make a meaningful contribution.

In short, you need to recognize when your work is ready to be published or whether it would be better to wait, or to communicate your progress in other ways, perhaps by presenting at a conference or by submitting a short communication or research letter. Today, a wide variety of channels are available for you to test the waters, to practice communicating to wider audiences, and to provide you with feedback that will help you further refine your work and your message.

WHAT'S NEW?

When you come to a point in your research where you are beginning to think about writing something up for publication, you need to look at your findings objectively and make a thoughtful appraisal of what you have to offer the research community and which journal and which format will work best to carry your research. This kind of objective assessment can be difficult, particularly for researchers just starting out in their career who are eager to publish and who may put a greater value on their results than the broader scientific community might. In this chapter, we look at how to assess your own work in terms of its originality and importance to the field and how to keep up-to-date with the latest information in your field.

When deciding what and when to write, you need to answer some simple but important questions, such as:

- What's new? Are the results or the techniques you used novel?
- Are your results complete? (Perhaps a submission as a rapid communication or research letter would be more appropriate.)
- Is your data set large? (A data paper maybe?)
- Will your findings contradict those of other researchers?
- Are your findings or conclusions going to be controversial?
- Do you provide or need to provide a detailed study of the literature on a particular topic, with a broad overview of other papers, with a new synthesis at the end or statistical analyses of the findings of others? (A review paper or a meta-analysis, respectively.)

This list could go on and on. The point is to be sure you are clear not only on whether you are ready to write but, if so, what type of paper you should write. These distinctions are crucial to success in narrowing your search for a suitable publishing venue (see chapter 5). Editors will be looking at your paper with the same key questions in mind. In an initial review of the manuscript, they will be able to fairly quickly determine whether the paper actually provides new information and whether it will be of interest to the readers of their journal. The Editors will also need to assess whether the findings, conclusions, or theories contribute to current understanding of the field. As we discuss in later chapters, several elements of your submission, including your cover letter, abstract, and conclusions, can work to support the main body of the manuscript to convince that first reviewing Editor that your science should go into peer review.

THE CHALLENGE OF KEEPING PACE

Making sure your science is novel is an increasing challenge. When young scientists write a dissertation, an exercise that is often their first serious attempt at scientific writing, they are usually required to write a literature review to display the breadth and depth of their knowledge of the field and to provide the background to the problem their research is trying to address. After writing their theses, however, some researchers become so absorbed in the daily demands of their jobs that they find it increasingly difficult to cope with the tsunami of information coming their way in journals, blogs, emails, and other professional communications.

In some areas of study, such as geology or physical anthropology, scientific progress takes place at a moderate pace and new insights arise over longer rather than shorter cycles of discovery and development. In other fields, however, particularly in the biosciences and biomedicine, groundbreaking discoveries happening at lightning speed. Researchers are developing new methods, generating new data, and gaining new insights faster than ever before. Researchers in these domains are in a constant race to keep up with the extraordinary pace of developments and to ensure that their work incorporates the most current information and is able to contribute meaningfully to the field.

A variety of factors can make it difficult for researchers to keep pace with scientific trends, particularly in fast-moving fields. One problem, of course, is simply time: there are only twenty-four hours in a day for reading, research, and family life. Another problem is that "traditional" access to newly published science, through university and public libraries, is becoming more restricted as libraries around the world cut subscriptions in response to increasing budget shortfalls. Yet another problem arises with the different technologies that are available for fast access and reading and their associated costs.

Researchers in remote areas and less developed nations face additional challenges. Lack of funding for research, equipment, and travel to conferences are all major issues, as are limitations in their access to the newest science due to financial restrictions, poorly stocked libraries, restricted or shared access to computers, frequent power outages, and narrow bandwidths, making downloading of larger files impossible. Some countries also actively engage in censorship, blocking access to sites deemed "unsuitable" for various reasons. Although several important initiatives developed over the past decade are making progress in mitigating inequities in access to scientific journals (see sidebar 3.1 on Research4Life), a great deal more needs to be done to help authors in developing countries to compete in the high-pressure arena that is modern-day scientific publishing.

SIDEBAR 3.1

Research4Life: Providing access to developing countries

RICHARD GEDYE
Director, Publishing Outreach Programs, International Association of Scientific, Technical and Medical Publishers (STM)

Research4Life is the collective name for four partnerships whose aim is to reduce the scientific knowledge gap between industrialized countries and the developing world. These partnerships provide countries across the developing world with access to critical scientific research literature in health, agriculture, and the environment. Together, these give researchers at over 6,000 institutions in 106 developing countries free or low-cost access to more than 8,500 of the world's leading science journals and online books, as well as to databases, indexes, references, and non-English resources. The number of institutions with access to Research4Life resources is increasing every year. Research4Life includes:

- *Health InterNetwork Access to Research Initiative (HINARI), which provides access to international research journals in medicine, nursing, and related health and social sciences. For more information go to www.who.int/hinari*
- *Access to Global Online Research in Agriculture (AGORA), which provides access to international research journals covering agriculture, fisheries, food, nutrition, veterinary science, and related biological, environmental, and social sciences. For more information go to www.aginternetwork.org*
- *Online Access to Research in the Environment (OARE), which provides access to scientific journals and other information resources in a wide range of disciplines related to the natural environment, from botany to zoology. For more information go to www.oaresciences.org*
- *Access to Research for Development and Innovation (ARDI), which provides access to international research journals in the applied sciences and technology. For more information go to www. wipo.int/ardi*

Developing countries and territories that qualify for access to resources are divided into two "bands," according to their national income. Beneficiary institutions in eligible countries include universities and colleges, research institutes, professional schools, extension centers, government offices, local nongovernmental organizations, hospitals, and national libraries. The Research4Life programs constitute a public–private partnership between the World Health Organization, Food and Agriculture Organization, United Nations Environment Programme, World Intellectual Property Organization, Cornell and Yale Universities, more than 190 science publishers led by the International Association of STM Publishers, and technology partner Microsoft.

For more information about the Research4Life effort and eligibility, see www.research4life.org.

Another reason that scientists have difficulty keeping up with new developments in their field is the sheer volume of information being published both in print and online. Every year, over 200 new journals are launched worldwide and the number of papers being published by existing journals is also growing rapidly. PubMed, the freely available index of biomedical abstracts (http://www.ncbi.nlm.nih.gov/pubmed) now includes over 20 million citations, and submissions have been increasing at a rate of about 4%

annually over the past twenty years (Lu, 2011). Authors must also keep up with the ever-growing number of channels through which information is being distributed. The Internet, cellular communication, new reading technologies, social media, and the latest application tools are continually opening up new ways of sharing information.

Part of your work as a scientist is to identify these new systems and technologies, to keep pace with the developments that are critical to your work, and to participate in the discussions and debates that they engender. It's also important to note that in this age of proliferation of communication channels, it is probable that by the time you've gotten used to using one set of channels, new ones will already have been developed and will be in common use. At the same time, new technologies and social networks are developing that can help you keep pace (see sidebar 12.1, "The Evolving Role of Mobile Apps in STM Publishing").

KEEPING UP WITH THE LITERATURE

Researchers have always tried to keep up with the current literature and developments in their field by subscribing to publications or by accessing them through institutional or public libraries. However, these days, pertinent new information is also found in a number of other formats and venues. For example, many publications contain not only peer-reviewed research papers but additional sections such as letters, rapid communications, reviews, news, opinion pieces and editorials, and special supplements. All these sections carry important information that reflects new trends and cutting-edge theories and which you therefore also need to monitor.

One way of keeping up with the latest science is to scour publication databases to find abstracts or, where available, full text articles that have recently been published in scholarly journals, particularly open access journals. These databases are supported by a wide range of providers, from academic institutions to for-profit publishers, and some are quite costly to access. Appendix 2 lists some of the most useful and popular free databases and search engines for scientific literature, including Google Scholar, PubMed Central, and the Directory of Open Access Journals (DOAJ). You can use these resources to find citations and abstracts to published papers, and sometimes also to access the full text of the article you are interested in. Other subscription-based resources are available to those who can pay to search for and access peer-reviewed journal articles. Some of these resources (e.g., the ISI Web of Knowledge) are widely used by researchers at universities and other research institutions that can afford the subscription price. Other appendices provide additional information about resources that can help you to locate papers, hunt for metadata to help you link to publications, and search for other ways to access published content. In addition to free databases of peer-reviewed and gray literature, publishers are also developing tools to help researchers navigate through the increasingly complex network of places where scientific information is produced and stored. ("Gray literature" refers to publications such as government reports, white papers, and preprints, that are often considered credible sources of scientific information but are not published by traditional publishers and are often not easy to find.) One way or the other, when you are considering writing a paper, one of your first tasks is to thoroughly search all the information resources available to you in order to ensure that you are presenting your work in the context of the newest information and ideas.

DOIs, HOW TO USE THEM, AND WHY THEY MATTER

One of the newer and more important tools that publishers have developed to help researchers more easily find electronic content is the digital object identifier (DOI). For the purposes of scientific publishing in much of the Western world, DOIs are administered by the organization CrossRef (www.crossref.org), an official registration agency of the International DOI Foundation (www.doi.org). CrossRef, launched in 2000 as a joint effort among academic publishers, is a not-for-profit organization intended to ensure the availability of persistent citation linking in online academic journals over time. Thanks to this system, even if a journal ceases publication, changes its web address, or updates its servers, papers in that journal can still be accessed via their DOIs.

In the early days of the Internet, publishers would put online versions of the articles they published on their servers, so each article would be associated with a URL. However, over time, servers had to be changed and updated, so the URLs would change as well. Authors who had bookmarked a URL as a link to the online version of a particular article would search for the URL and find that it no longer existed. DOIs solve this problem because each DOI will always be associated with a particular article; if a publisher changes servers, or a publishing company change hands, the responsible parties only need to tell CrossRef. CrossRef will then update the information in their database, so that the DOI link remains active and accurate.

In simple terms, a DOI is a unique string of numbers and letters associated with a specific piece of digital information. A DOI can be assigned to any digital object: a scientific article, a data set, a map, an image, a specific tabular display of information, or a piece of music. You probably now regularly see DOIs in very small print somewhere near the beginning of each article (see sidebar 3.2). If you have some information about a journal article, such as part or all of a bibliographic citation or just the DOI, you can go to the CrossRef website to look up the full citation for the article. DOIs link directly to the article itself if it is in an open-access publication or to whatever information the publisher will allow you to see for free, together with instructions on how you can purchase a copy of the full paper.

SIDEBAR 3.2

What authors need to know about CrossRef DOIs, CrossCheck, and CrossMark

CAROL ANNE MEYER
Business Development and Marketing,
CrossRef

What if you could ensure that the references in the papers you submit for publication are accurate? Or check that when someone cites one of your papers the reader can always follow a link from that citation to your paper? Or make sure that papers published by other authors do not copy (either knowingly or by mistake) any of your own writings? Or be certain that the research you rely on hasn't changed since you originally read a particular paper? CrossRef, a not-for-profit association of more than 1,000 scholarly publishers, provides services to make sure that research can be discovered, linked to, is original, and can be trusted.

REFERENCE LINKING AND DOIs

The best-known service CrossRef offers is persistent reference linking through the use of digital object identifiers (DOIs). Once a publisher accepts and publishes your paper, you may be thinking about your next big research project, but CrossRef and its member publishers want to make sure that readers can find your last paper for a very long time—in fact, forever. That's why CrossRef member publishers assign DOIs (persistent links) to the content they publish. That way, even if the journal your article is published in changes its URL addresses or gets sold to another publisher, or even ceases publication entirely, a DOI link to it will still work and will still allow readers to access your paper.

You may see DOIs displayed in a few different formats. They may look like this: **http://dx.doi.org/10.1087/20110202***. In this case, all you have to do is put the URL in a browser or click on the link to go to the document. They might also look like this:* **DOI:10.1087/20110202** *or even like this:* **10.1087/20110202***. Sometimes, they may be hidden in a reference list, behind a link that says "CrossRef" or "Full Text." Regardless of how a DOI looks, you can always "resolve" it—*

that is, it will take you to the landing page of the document—by putting it after http://dx.doi.org in the address bar of your browser.

If you ever find a DOI link that doesn't work, first check that you didn't make typing errors or transpose a letter or a number (lowercase el (l) for the number one (1) or the the capital oh (O) for the number zero (0)). To avoid these kinds of mistakes, copy and paste DOIs instead of typing them whenever you can. If that still doesn't work, please report the broken DOI (http://www.crossref.org/DOI Complaint) so that CrossRef can contact the publisher and get it fixed. Since CrossRef DOIs are all about persistent links, we take broken DOIs seriously.

You can also play a role in ensuring the persistence of scholarly content—both your own and that of authors you cite. When you prepare your references for publication, make sure to look up and include DOIs for references in your citation list. CrossRef has several tools to help you do this, and they don't cost individual researchers anything. One is the Guest Query form (http://www.crossref.org/guest query), which allows you to type in the bibliographic information (author, title, journal, volume, issue, page number) for your reference and get back its DOI. Another is the Simple Text Query form (http://www.crossref.org/SimpleTextQuery), into which you copy a whole reference list, submit it, and get back a list of DOIs.

You should always use the DOIs for your own articles and those of other researchers whenever you reference them, whether on your own home page, in your CV, or in blogs or Twitter posts. Using DOIs is particularly important because publishers are starting to collect article-level metrics from the web to show where individual articles have been referenced. If you don't use the DOI to reference content, these metrics can't identify the web citations.

PREVENTING PLAGIARISM

What about plagiarism? Authors should always carefully attribute any work of others they use in their writing, but unfortunately, not all authors are as diligent, or as ethical, as they should be. To discover cases of plagiarism, many scholarly publishers now use CrossCheck, powered by iThenticate, which is software that compares submitted manuscripts with millions of published scholarly and web documents. CrossCheck reports on the percentage of similarity and flags passages that are a match. This screening process protects your published content against those who may copy it, and it helps to prevent cases of plagiarism in published journals that can lead to embarrassment at best and ruined careers at worse.

IDENTIFYING RELIABLE CONTENT ON THE WEB

In its efforts to increase the trustworthiness of published scholarly research, CrossRef has recently developed another service, called CrossMark, that helps researchers identify whether any of the content they find on the web has had significant updates since it was originally published. When you see a document that contains the CrossMark logo you will be able to tell that the publisher has committed to keeping it updated. You will also be able to click on the CrossMark logo to get up-to-date status and publication record information about the document to determine if it is current or whether it has been corrected or even retracted. Clicking on the CrossMark logo could save you the embarrassment of citing research that may have changed since you originally looked at it. It will even work on PDF files that you download to your computer or save in a database.

For more information on CrossRef, CrossCheck, and CrossMark, you can visit these websites:

- Information for Researchers: http://www.crossref.org/05researchers/index.html
- Guest Query: http://www.crossref.org/guestquery
- Simple Text Query: http://www.crossref.org/SimpleTextQuery
- DOI Complaint: http://www.crossref.org/DOIComplaint
- CrossCheck and Researchers: http://www.crossref.org/crosscheck/crosscheck_for_researchers.html
- CrossMark: http://www.crossref.org/crossmark

The same idea—assigning unique identifiers to certain kinds of objects—is also being applied more broadly within the scientific community. For example, the Open Researcher & Contributor ID (ORCID) Initiative (www.orcid.org) assigns unique digital identifiers to individual scientists and their research, which will help you to identify and track other researchers who are doing work similar to your own. Identifiers are also now being assigned to data sets, as described in sidebar 3.3.

SIDEBAR 3.3

DataCite: linking research to data sets and content

JAMES L. MULLINS
Dean of Libraries and Esther Ellis Norton Professor, Purdue University

Researchers have long wanted a system that would allow greater access to each others' data sets, as well as some means of archiving data so that it can be shared and reused in future studies. The need for a process that would provide reliable, consistent, and perpetual access to data further increased following a mandate by funding agencies in the late 1990s that required researchers to make data sets available for other users. In response to this need, DataCite was founded in 2009, specifically for the purpose of managing, sharing, and preserving data sets. The DataCite system is based on Digital Object Identifiers (see sidebar 3.2), and uses the same principle of applying a unique identifier to data sets associated with published papers, although in this case DataCite, not CrossRef, works in association with a number of research libraries around the world to issue the identifiers.

WHY CITE DATA?

Data should be cited in the same manner that articles and books are cited. Citation enables data to be verified and reused, and allows the impact of data sets to be tracked as a means of recognizing and rewarding researchers for their contributions. As a result, impact factors are now being calculated for data sets.

WHAT IS DATACITE?

DataCite began in 2006 when Dr. Jan Brase of the Technische Informationsbibliothek Hannover (the German National Library of Science and Technology, TIB) recognized the need for a consistent method by which data sets could be tagged with a unique identifier to ensure accessibility and retrievability. In response to this idea, representatives from six countries came together in London in 2009 to formally inaugurate DataCite as an international not-for-profit organization. The founding members included the British Library, the Technical Information Center of Denmark, TU Delft Library, the National Research Council's Canada Institute for Scientific and Technical Information, California Digital Library, Purdue University Libraries, and the German National Library of Science and Technology. Since then, libraries from Australia, France, Sweden, and Switzerland have also become members and take responsibility for developing services for registering DOIs for data sets in their own country.

HOW DOES A RESEARCHER OBTAIN A DOI FOR A DATA SET?

Researchers or organizations can work with a DataCite member in their region to obtain DOIs for data sets (see www.datacite.org for a complete set of members). In the US, for example, there are three DataCite members: California Digital Library (CDL), which is part of the University of California, the Office of Scientific and Technical Information in the Department of Energy, and Purdue University Libraries. These three organizations establish registration policies and procedures for anyone in the US who wants to assign DOIs to data sets. CDL and the Purdue University Libraries collaborated on the development of Easy ID (EZID), a program that creates, manages, and stores DOIs and the associated metadata for subscribers.

WHAT ARE THE LONG-TERM IMPLICATIONS FOR DATA THROUGH DATACITE?

Once a DOI has been assigned to a data set, the associated metadata is registered and will identify both the creator and the content. Although the DOI is assigned by one of the DataCite members, the researcher is responsible for depositing the data in a trusted repository for long-term preservation. However, even if the data set does not survive, there will still be a record of its existence and the associated metadata through the registration of the DOI.

FOR MORE INFORMATION

For information on the metadata schema for DataCite, see Starr and Gastl, 2011. See also the following websites:

- DataCite: http://www.datacite.org
- California Digital Library EZID: http://www.cdlib.org/services/uc3/ezid
- Purdue University Libraries Distributed Data Curation Center (D2C2): http://d2c2.lib.purdue.edu.

KEEPING UP WITH CONVERSATIONS

In addition to communicating scientific progress through print and online publications, researchers have long relied on professional meetings and conferences for face-to-face exchange and discussion about research in progress. A long-standing tradition of scientific communities since professional scientific societies first formed in the mid-1600s has been to gather in person to share information, network, and debate the latest ideas and research results.

In the pre-Internet days, professional meetings were perhaps the most effective way that researchers could engage with their professional community and gain recognition from others. Certainly, participating in professional meetings provides you with the opportunity for direct discussion with other experts, and with a way to learn who's who and to see firsthand how senior members of the community present both preliminary and complete findings. By attending and presenting at meetings, particularly through a less formal medium such as a poster presentation or round-table discussion, you can gain experience in presenting your material and get yourself known by peers and senior scientists. All of these activities will stand you in good stead as you progress up the scientific career ladder.

Discussing your work face-to-face with other researchers is a critical part of becoming an effective scientist. Many scholars in academia make a point of bringing their students and research lab members to scientific meetings for the purpose of introducing them to others and to mentor them on how to make the most of this kind of professional gathering. However, as important as such conferences are, and will probably continue to be for a long time, direct face-to-face meetings are no longer the only way that researchers can discuss issues and interact in real time. The Internet now allows you to communicate remotely through video and audio links, share text and images instantaneously and at no cost, and in doing so lets you participate in an online community. These virtual communities are growing in importance and are becoming critical resources for communication and debate among scientists.

Journal publishers and others are working to help researchers keep up with all the new information and online communities being formed through the Internet. For example, many publishers are enhancing their journal websites by offering users the opportunity to have information sent (or "pushed") to them in regular updates or by packaging information in ways that will entice (or "pull") users to revisit their websites frequently. You can access new content by signing up for electronic alerts, RSS feeds, podcasts, press releases, mobile notifications, or other publication-related news.

Publishers that have sufficient resources—often the large for-profit publishers—are able to keep the information on their primary websites fresh by promoting new and supplementary content from their publications and affiliated societies, and by allowing users to link to useful portals and databases, blogs, newsletters, and other special features. Some publishers are also trying to help researchers identify trends in their research community by offering innovative networking and search tools, such as Nature Publishing Group's Nature Network (network.nature.com), and the American Institute for Physics' UniPHY (www.aipuniphy.org/Portal/Portal.aspx).

Professional societies for Editors and publishers are also trying to help their members to stay current with new topics and trends. For example, organizations such as the International Committee of Medical Journal Editors (ICMJE, see www.ICMJE.org) or the previously mentioned Council of Science Editors serve as independent industry watchdogs in ways that for-profit publishers cannot. The ICMJE, for example, has developed policy statements, guidelines, and other materials that publishers, Editors, and authors can look to for guidance on various topics, including ethical considerations (e.g., work with human and animal subjects, and conflicts of interest) as well as authorship and peer review.

Blogs, Twitter feeds, social bookmarking services, and other online resources allow users to easily recommend papers and share information and opinions. However, becoming aware of and participating in these exchanges remains an ongoing challenge for researchers in the developing world. Researchers, young and old, will need to keep their ears open and their conversations active in the community of their intellectual peers. The job of Editors is to filter the new from the old. As an author, part of your job is to convince Editors that you are ahead of the field and that your science will make a worthwhile contribution to their journals.

THE BOTTOM LINE

Among the key skills that authors must develop are the ability to recognize when a piece of research is far enough advanced to be worthy of publication and what kind of paper will be the best vehicle for the information. Ensuring that each manuscript contains some new information is also critical, but to do that authors need to keep up with all the latest developments in their field, which can be a daunting task. Monitoring both traditional venues (e.g., conferences, journals) and new outlets (e.g., blogs, alerts, RSS feeds) can help authors to keep up-to-date with the newest developments. Authors often want to know how they can begin writing their manuscripts. We don't go into that topic in this book, not only because there are many good texts available that address this questions (see appendix 1), but also because if your study is suitably conceptualized, well designed, and properly carried out, your findings will be new and exciting and your data will tell you how to start your writing.

CHAPTER FOUR

Authorship issues

Trust is among the fundamental bases on which scientific communication rests: trust that the authors have fairly and accurately reported their findings and disclosed all pertinent commercial and professional relationships that could bias those findings, and trust that Editors have exercised sufficient diligence and skepticism to ensure accurate reporting and disclosure by authors.

COUNCIL OF SCIENCE EDITORS, White Paper on Promoting Integrity in Scientific Journal Publications, 2009

Authorship is a vital part of a scientific career because it is one of the main benchmarks by which scientists judge one another. Your publication list is one of only a very small number of criteria by which you will be judged when it comes to getting a job, a grant, tenure, promotion, or any other milestone in a successful career in science.

As a result of the pressure to publish, authorship can be an extremely sensitive subject and a common cause of disagreements among researchers—sometimes leading to long-standing hostilities. Many a successful scientific partnership has ended as a result of a dispute over authorship. These quarrels can have a number of different causes. For example, a scientist may want to be first author, or to be placed further forward in the author list, but the other authors may disagree. Sometimes, researchers are left off the author list, even though they feel they deserve to be on it. Occasionally, people's names are added to the author list without their consent or knowledge, and they don't find out until after the paper has been published. Senior scientists have been known to insist on being on the author lists of papers written by their junior staff and students, or junior scientists are left off the author list, despite having done a major share of the work.

Merriam-Webster's dictionary defines "author" as "one that originates or creates" and "the writer of a literary work," while oxforddictionaries .com speaks of "a writer of a book, article, or report" and "an originator or creator of something, especially a plan or idea." All these definitions describe not only the action of writing a text but also of generating ideas, planning, or contributing in some other way. In science, the meaning of the word "authorship" stretches even further, to include all sorts of additional contributions that can lead to a place on that all-important author list at the top of a scientific paper. The uncertainty about what a person has to do to be counted as an author has been the cause of many problems. Certainly, if authors have done one of the following, they'll be on the author list:

- written the paper
- designed the experiments
- gathered the data
- analyzed the data
- contributed original idea(s) to the research
- obtained the grant that supported the work.

But what about the following contributions, all of which are close to the borderline between named authorship and a grateful mention in the acknowledgments? What if an individual:

- translated the paper into English?
- substantially improved the English in the paper (originally written by a nonnative English speaker)?
- took a poor-quality paper and greatly improved it by changing its structure or clarifying the text?
- gathered some of the data—how much?
- carried out some of the analysis—how much?
- contributed one original idea—how important?

In short, the line between authorship and acknowledged contributor can be very fuzzy, and can therefore easily be the cause of disagreements and disputes (see also sidebar 4.1).

ORDER MATTERS

There are recognized conventions that govern author order, but these vary greatly depending on the subject area (Brereton 2010). In some fields (mathematics, particle physics, astronomy), authors are often ranked in alphabetical order, which prevents a lot of arguments. However, in other areas, such as the natural sciences and biomedicine, the order in which names appear at the top of the paper usually depends on the person's contribution to the paper itself or to the research on which the article is based. The first name on the list is usually the person who did most of the work. The succeeding names are then ranked according to their level of contribution. The last name on the list in a multiauthor paper is generally understood to be the leader of the lab group or the Director of the department or institute in which the research took place. This senior figure may not have played any active part in the research but probably had to approve the final version of the paper and may have provided advice or guidance, read through one or more of the earlier drafts, ran the facilities in which the work was carried out, or possibly obtained the grant that supported the research.

In the past, many papers had just a single author or perhaps two or three at most. Today, however, we are in the era of the multidisciplinary, multi-institution, multiauthor study, as scientists from different specialties, institutions, and countries pool their knowledge and expertise to address complex scientific and medical problems. Sorting out who goes in what order can be a major headache for collaborators, as well as for Editors, who must find space on the printed pages to list all those author addresses. When you submit your paper, you need to be sure that the order in which the authors' names appear is exactly as agreed among all the coauthors and is as you wish them to appear in the final publication.

AUTHOR ORDER CONVENTIONS

In scientific and medical publishing, the first author position is valued most highly, and not only because the person listed first is assumed to have done the most work. First author is the prime position and is given the most weight when your publications list is being evaluated by an employer or grant-awarding body. First authorship also provides the greatest level of visibility. When your paper is cited by journals that use "*et al.*" (meaning "and others") which many publications do for papers with three or more authors, all but those first few names become invisible. In the text, citations to papers with three or more authors usually appear as only the first author (e.g., Jones *et al.*, 2010)," even in publications where the first three or four names are listed in the references section (also called the "literature cited" or "citations" section). When a two-author paper is cited, both authors are named in the text.

When two scientists undertake a research project, author order will be based on who did the greater amount of work. However, issues of order are less straightforward when the two researchers share the work and the responsibility more or less equally. Both need the boost to their reputation that comes from being first author, so what should they do? Someone has to go second. One way that authors have solved this dilemma is to ask the journal to place a footnote somewhere in the paper (either on the first page or in the acknowledgements section), stating that they both contributed equally to the work.

Other problems can arise, for instance when a senior scientist in a position of authority insists on a place in the author list without having contributed anything much to the paper. Alternatively, an author may offer an "honorary" or "guest" authorship to someone they feel will add weight to the publication, even though that individual made little or no contribution to the work (see chapter 11). Finally, difficulty in assigning author order for a multiauthor paper also arises when each individual had a variety of different roles and responsibilities. Weltzin *et al.* (2006) suggested a way of dealing with some of these issues, namely to include, with each paper published, a textbox that lists the initials of each author followed by a brief description of what they contributed to the paper (see box 4.1). Not many journals have actually taken up this idea, but in January 2010, three major journals, *Science*, *Nature*, and *Proceedings of the National Academy of Sciences*, announced the introduction of a new policy (Alberts 2010), in which authors are asked to describe their contributions to the paper before it can be accepted. In addition, the original data on which the paper is based must be verified by a senior author to ensure that the manuscript correctly represents the data. Where more than one lab or institution is involved, a representative from each group is asked to carry out this verification. In this way, the Editors of these three top-ranking journals hope to prevent the addition of "honorary authors" and to minimize the risk of various types of scientific fraud.

EARLY AGREEMENT CAN SAVE UNPLEASANTNESS LATER

One of the best ways for a group of researchers to minimize the risk of having arguments over authorship is to agree at the very beginning what will

Author contributions to this article

JFW co-conceived and co-developed the idea for the manuscript, co-refined the intellectual content and scope, edited all drafts, prepared the final version of the manuscript, and facilitated the gathering of contributors. RTB co-conceived and co-developed the idea, edited all drafts, and assessed historic trends in authorship in ecology. LTW initiated the project, co-developed and co-refined the intellectual content, and wrote the first two drafts. JKK co-developed the idea, edited all drafts, and conducted the keyword search. ECE co-developed the idea and coordinated the authorship survey. JFW is the guarantor for the integrity of the article as a whole.

BOX 4.1 Example of a box acknowledging the contributions of each of the authors of a paper (from Weltzin *et al.* 2006, used with permission). The initials represent the authors, whose full names appear at the beginning of the article and therefore don't need to be repeated.

constitute authorship and what will dictate author order. At this early stage, no one can be absolutely certain about who will do the most work or who will carry out a particular task, but at least a baseline has been established. Circumstances may change and unexpected issues may arise, so that someone who had planned to carry out a particular part of the research may be prevented from doing so, while someone who had expected to play a relatively minor role may end up doing much of the work. Therefore, although the members of a research team should agree on a plan, including everyone's roles and responsibilities and the order in which authors' names will appear on any resulting papers, all involved should also understand and agree that the original order may need to change as the work progresses (see sidebar 4.1).

SIDEBAR 4.1

Honesty in authorship

MONICA BRADFORD
Executive Editor, *Science*

Although much has been written in the past twenty years about what qualifies an individual to be an author on a scientific publication, authorship disputes continue to occur. Training programs to teach graduate students research ethics have been in place in the US since the mid-1990s, and many journals routinely require researchers to clearly indicate their contribution to the work being presented in the article. All of these resources provide clear guidance on how to determine who is an author and who should be acknowledged. So how does it happen that an article is far along in the publication process before authorship disputes come to a head?

Clearly, the opportunity for preventing such disputes occurs long before a paper is submitted to a journal. Discussing authorship expectations from the start of the research project is a good way to avoid last minute disputes. Steneck (2010) also advises early-career scientists to study "the standards of research practice in their areas of investigation" before starting a project, or "run the risk of making mistakes and getting into difficult situations." But discussing roles and knowing the standards may not be enough to prevent conflicts, because research is increasingly multidisciplinary, requiring diverse contributions from individuals often working in different institutions and different countries. As part of a collaborative effort, each researcher brings not only knowledge and expertise but also cultural norms and expectations to the research process. As a result, coauthors often need to navigate complex human interactions over time and distance. Further complicating the matter, authors are far from being equals in terms of independence and career security. And authorship decisions can be inherently complex; more is involved than performing experiments and writing up the results. Strong interpersonal skills, open communication, and a supportive research environment are essential to transform a diverse group of individuals into a team of authors who trust each other and their research results.

Central to this trust is integrity: "intellectual honesty and accuracy in representing contributions to research" as well as "other facets of individual integrity—collegiality, communication, and sharing of resources" (Cho et al. 2006). These elements seem so fundamental to the practice of science, yet their absence is often the root cause of authorship problems. Steneck provides an interesting insight into the relationship between authorship and integrity when he notes that the most significant tests of integrity "are the dozens of routine decisions scientists make every day." By the time an authorship team starts to write up the research, the researchers have made multiple decisions, individually and jointly, that impact trust and research integrity. He goes on to point out that the consequences of small decisions to the overall integrity of the research are not immediately evident, "which makes it easier to justify bending rules and cutting corners" (Steneck 2010).

From the point of view of an Editor, it seems that researchers often face difficulties during the publication process because senior scientists have not taken the time to create an environment that fosters integrity. C. K. Gunsalus points out that the informal activities that "students absorb in the hallways, laboratories and hospitals" are as formative as the formal academic curriculum (Gunsalus 1997). She posed these questions to senior scientists: "What messages do students pick up about authorship and publication practices? How do they see mentors reconcile a desire for a hefty publication record with admonitions not to engage in 'salami science' or divide work into 'least publishable units?' ... Most important, do the rules apply to everyone in your environment or only to students?" (Gunsalus 1997). As in any endeavor, actions speak louder than words and mentors need to be particularly careful about the examples they set with regard to how they expect students and postdoctoral fellows to behave. When beginning collaborations with another research group, it is important to determine whether they share your expectations with regard to investigator behaviors.

In a research environment devoid of trust, valuable time that should be spent moving research forward is spent arbitrating disputes. In an environment devoid of integrity, the consequences can be significant. As Floyd Bloom asserts, "Authorship and collaboration problems are a serious threat to the research enterprise and to the motivation of young scientists, especially when they involve misap-

propriation of ideas and data" (Bloom 2000). In the most egregious cases, authors face institutional investigations and/or lawsuits and the public loses confidence in scientific results. It is only natural for a research team to want to be first, or to have their work featured in the press, or to change what is written in textbooks. But those outcomes will be temporary if decisions are made along the way that erode the integrity of the research. Researchers at all stages of their careers must be repeatedly reminded that "integrity and reputation are among our major assets as scientists" (Levy 1997). Once lost, it is almost impossible for a scientist to recover these critically important assets. So choose your coauthors carefully and don't shy away from explicit discussions among the team about protecting the integrity of your research at every step along the way to publication.

CHOOSING COAUTHORS

As mentioned earlier, people reviewing your publications list will tend to put the greatest value on those papers for which you were the sole or first author, particularly when they are considering whether to offer you a job or a promotion. However, few if any researchers can work entirely alone and, paradoxically, multidisciplinary (and therefore by definition multi-author) papers are widely recognized as vital in tackling many of the most intractable problems we face today. The question then becomes how do you find the right coauthors to share the workload as you pursue your research goals, and who will be able to make the most valuable contributions as you prepare to publish the results.

Of course, you may not need to look further than your own research group or departmental members, but you can also look further afield. Obvious choices for possible collaborators are colleagues working on similar research topics from other departments in your institution or from other universities or research institutes. Consider approaching the author of a paper that you have read and admired, or a speaker at a conference you attended, if you feel they could bring valuable knowledge or a new perspective to your work. You may also want to consider a researcher from another discipline, with different skills or knowledge, or someone familiar with a particular technique that you have no experience in (or someone who has access to the equipment required to carry out that technique). Obviously, before you invite a complete stranger to be a coauthor on your paper, you will want to find out a bit more about that person, so you will need to do some research. But if you are fairly confident that a particular individual could make a valuable contribution, don't be shy about asking them—many researchers will be pleased and flattered to be approached, and the very worst that can happen is that they say "no." And of course, you must never add anyone's name to an author list without their knowledge or permission.

THE PROS AND CONS OF HAVING COAUTHORS

Provided you are working with the right people, having coauthors can definitely strengthen a paper. Having someone to share the workload and to discuss options and ideas with can also be a huge help.

You may also want to consider inviting a big-name scientist to join the author list, but only if you think the person will provide meaningful input. You should not offer authorship as a "gift," and hopefully an experienced senior researcher would refuse anyway. In cases where the individual is your supervisor or the head of your lab, this may be an obvious choice or necessity. One way or another, the possibility of adding a well-known name to your author list is a difficult decision that junior authors have always had to wrestle with. Having a big name as a coauthor may bring extra weight and prestige to a paper and the sight of a well-respected name somewhere in the author lineup may be just enough to save a borderline paper from the reject-without-review pile. However, that same big-name presence may divert some of the attention away from you, the junior author, at a time when you are trying to develop your own reputation and so need all the recognition you can get.

Leimu *et al.* (2008) wanted to know whether having a big-name author (whom they called a "bigwig") on the paper would increase the number of citations, regardless of whether the increased citations were because this senior author improved the quality of the paper or was attracting extra citations just by being there. They looked at papers in a midranking ecology journal and used the *h* index (Hirsch 2005) to measure the senior author's scientific stature. They found that for papers with up to three authors, having a well-respected senior author as a coauthor did lead to higher citation rates, but this effect disappeared when there were four or more authors. This result led to a published discussion with Havens (2008) regarding whether the effect was indeed due to the extra quality that such an author would bring to the work or whether senior authors were only likely to become involved if the work was of particularly good quality to begin with.

NOTE FOR NONNATIVE SPEAKERS OF ENGLISH

Authors whose primary language is not English are often strongly encouraged to seek help with the English in their papers (see sidebar 4.2 on some of the challenges faced by international authors and 8.3 on how to choose a good language-polishing service). One option mentioned elsewhere in this book, and in countless Editors' letters to authors, is to seek the help of a colleague or acquaintance (or even a scientist whom you have never met, but who is a native English speaker and an expert in your field) who can work with you to improve the English in the text. If you find yourself seeking this kind of help on one of your papers, you need to make it very clear, right from the beginning and preferably in writing, what you are offering in return. In most cases, a mention in the acknowledgments should be enough. Unless the individual makes a substantial contribution to the *intellectual* content of the paper, this type of language assistance does not necessarily warrant inclusion as an author. However, there have been occasions when this was not agreed at the outset, leading to arguments at a later stage.

SIDEBAR 4.2

The challenges of publishing as an international author

KEPING MA
Professor, Institute of Botany, Chinese Academy of Sciences

The basic goal of science is to add to the world's collective knowledge and this goal compels us to develop and do good science. Scientific publication, in turn, is how we confirm and refine new discoveries to determine what really is new knowledge. International scientists (that is, those working in countries outside those where the majority of science is being published) face particular challenges in publishing, and those for whom English is not a primary language face even greater constraints. The challenges facing nonnative speakers and writers of English are rooted not only in linguistic differences but perhaps more importantly in cultural differences and in the accepted ways of presenting papers for publication.

International authors begin by facing exactly the same challenges as any other authors. First and foremost, they have to do good science, following international standards on research integrity. Once their science is done, they have to select an appropriate journal for their work, write clearly according to standard formats for presenting science, and follow the instructions to authors exactly. However, inexperienced international authors may face even greater challenges than those of their Western counterparts. For example, international authors may not be able to get advice as easily from senior scientists, in part because of cultural differ-

ences that make it difficult to ask for this kind of help. Some international authors may have limited Internet access, may have difficulty getting copies of back issues of journals to study as possible targets for submission, and may have limited financial resources to pay for page charges or open access fees. These constraints can hamper the ability of international authors to identify the most appropriate journals to submit to.

Once an international author has selected a journal, other challenges arise. For example, if a journal suggests writing a cover letter, nonnative speakers of English may again be at a disadvantage, in that they may have limited knowledge of how to write any kind of formal letter in English, much less a cover letter to an Editor who will be making critical decisions about their paper. Some journals may ask authors to suggest reviewers, another possible cultural challenge for international authors who may be unfamiliar with this kind of practice and may also be hesitant to recommend someone they don't know personally.

If an international author does get a paper into review and must then respond to a decision letter, replying to criticisms may also present unique difficulties. For example, in some cultures, authors may be reluctant to publically point out that an expert reviewer misinterpreted something. Without the support of an experienced mentor, a young international author may believe that all the reviewers' comments must be followed, even if they are not completely correct. Of course, nonnative writers of English face double difficulties when writing in English, both because they are writing about something highly technical and because they are writing in a language that is not their native tongue. Editors and authors recognize that there is often at least some bias in the peer reviewers' reports of papers written by nonnative writers of English. Whether reviewers exercise this bias consciously or unconsciously, this prejudice makes it difficult for international authors to compete on a level playing field with their peers who can write English with the confidence of a native speaker.

Despite these difficulties, international authors are submitting their work to high-ranking journals in ever greater numbers. Journals, in turn, are paying more attention to the needs of international authors; many are trying to provide resources that address the particular needs of scientists who are doing excellent science and want to share it with their colleagues around the world.

WHO GOES IN THE ACKNOWLEDGMENTS?

If someone has helped you with any aspect of the paper (though not enough to have earned a place on the author list), use the acknowledgments section to thank them. You must also remember to thank the funding agency that gave you financial support—they may not give you any more money if you don't offer this simple courtesy. The following list shows some of the many possible contributions that could earn various individuals a place in this section of your paper. They may have:

- provided technical assistance, under your direction
- allowed you to use their graphic illustrations or photos
- given you some material for use in your research
- discussed your ideas with you and provided advice or suggestions
- read early drafts of the paper and made useful comments
- been the subject(s) of your study, answered a questionnaire, or provided other information.

Remember, it can never hurt to say "thank you," but forgetting to do so could certainly lead to problems later on.

THE ROLE OF CORRESPONDING AUTHOR

As a sole author, you will most likely be doing all the work associated with getting the manuscript published. However, when a group of authors collaborate on a project, at some stage they need to decide who will be the corresponding author (CA). The CA is not necessarily the first named author on the list, although this is often the case. In fact, the role of CA can be played by anyone on the author list. Among a group of scientists who are all nonnative speakers of English, the CA may be the person who speaks the best English. Alternatively, the scientist who has had the most experience with the publishing process may be the best person to take on this role, or it could be the most senior or the most junior researcher in the group.

So, what does the CA do? From the time of submission, and to some extent before that, the CA's job is to carry out various tasks associated with the manuscript as it moves through the peer review and publication process. In addition, no matter how many authors are associated with the paper, the CA is always the "point of contact," the organizer, and the representative of the author group in all dealings with the journal. The CA is also responsible for making sure all the authors have read and signed off on the final version of the paper before submission.

Ideally, Editors and editorial staff prefer to deal with only one person on the author list of each paper. The CA's name and all contact details should therefore appear separately, on the first page of the manuscript, or will need to be provided in a designated section in the journal's online submission form. The CA's duties usually begin at an early stage and may include orchestrating the different phases of manuscript preparation, correction, and approval by all the authors. The CA may also be delegated to research various journal criteria to see which publication is the best match for the paper. However, the main work begins when the paper is sent off to the chosen journal. In short, the CA's job is to relay every communication from the journal to all the other authors, and to gather and collate all the contributions from those authors and send the final result back to the journal.

As CA, you will probably be the person who uploads the manuscript and all the accompanying figures, appendices, and other materials to the online submission system, or the one who packages and posts the required number of copies of everything to the journal offices. The decision letter, when it finally arrives, will come to you, and it is your responsibility to relay copies to all your co-authors as quickly as possible.

If the decision was to reject the paper (with or without review), the group will need to decide where to send the manuscript next, which may involve you in further research into journal requirements. If the decision was a request for major or minor revisions, the authors will have to move into a new phase of discussion regarding how to address all the recommendations and comments (see chapter 10). In any case, the CA may be chosen to do much of the revision work on the text, to circulate drafts, and to incorporate everyone's corrections before sending it off to the same or another journal.

In cases of major or minor revisions, journals will usually give authors a deadline by which revised text must be submitted. This may be three months, six months, or more. As CA, it will be your job to find out what the time frame is and to ensure that all the authors provide their input within a reasonable period. If the deadline passes and the revised text has not been submitted to the journal, the Editors may decide to "close the file" and archive the paper. If a further draft arrives after the deadline, it will often be treated as a new submission and the whole review process may have to be started again from the beginning. Getting multiple authors to submit revisions in a timely fashion can require a lot of tact, as well as equal amounts of begging and bullying. If a coauthor is being very slow, you may need to ask the most senior author or other authors in the group to apply some pressure. At some stage in the process, the journal may require signed copyright forms from all the authors—you will be expected to distribute these forms to your coauthors and to make sure everyone gets them back to you or sends them direct to the journal. There may also be author billing forms, reprint order forms, and other paperwork to deal with.

If the paper includes figures or tables that have appeared in another publication, be it a book, journal, magazine, newspaper, or website, the Editor will also want to see written permission to use the material from that other publication. Permissions offices can be notoriously slow to respond to these requests and may take many weeks or even months to reply, so start this process as early as possible.

Later, when the proofs arrive, the CA will also need to distribute the materials to all coauthors, and then collate suggested corrections and return the manuscript to the journal. Finally, if at any time the journal staff have questions, it is the CA they will be contacting for the answers.

Being CA can involve a lot of extra work and can make you quite unpopular with your coauthors, as you will constantly be asking them to send back signed forms, return their corrections, and other materials. However, on the positive side, you will gain a new understanding of what to do at each stage of the publishing process, which will stand you in good stead when you are trying to get your next paper published.

FOREIGN AUTHORS' NAMES

One other source of confusion associated with authorship deals with the names of foreign authors. Editors sometimes have great difficulty in understanding and correctly reproducing, archiving, and indexing foreign authors' names, particularly when they are from countries that do not use the Latin (Roman) alphabet or have different naming conventions than those with which the Editor is familiar.

Chinese names are a particular case in point. In China, the usual practice is to put the family name before the personal name (so that John Smith would be called Smith John). Editorial staff who are unaware of this practice, and who also cannot distinguish between first and last Chinese names, often make the mistake of entering such names in the wrong order in the author database. To further complicate matters, some Chinese authors are aware of this problem and therefore enter their names into the manuscript tracking system in the Western style, with the personal name first, only to have the Editor, in an attempt to enter the information correctly, reversing the personal and family names back again! The next time those particular authors submit a paper to the same journal, they may represent their names a different way. As a result, it is quite common to find the same author listed in two or three different ways in the database. Such errors can occur in the case of authors from any country or culture where names do not follow the English conventions of order or spelling. Hungarians, for example, also place the family name before the given name, and authors from Brazil often have multiple family names. The terms that denote "from," "son of" (e.g., von, d', de, bin) and similar relational information can also lead to duplication and other problems in the author database.

If you believe that your name might cause some confusion among editorial staff, you should consider explaining the format of your name in a "notes" section of the manuscript submission system or even in your cover letter. If you are the CA, it will be your responsibility to make sure that all the names of your coauthors are correctly transmitted to the editorial office and that given names and family names are clearly indicated. Above all, however, you need to use the same format of your name every time you submit a paper to an English-language journal, no matter what your nationality. If you publish a paper as Jane Smith and then another as Jane B. Smith, you could very well end up appearing twice in the index.

THE BOTTOM LINE

Scientists can be very sensitive about issues of authorship. Agree on author order as well as roles and responsibilities at an early stage of the process. Think carefully about other individuals who have contributed to the paper to a lesser degree and be sure to include them in the acknowledgments. Some forethought and consideration can avoid unpleasant disputes; your early experiences as an author or CA will lead to a better understanding of the publishing process—which will make it seem a lot easier the next time around.

CHAPTER FIVE

Choosing the right journal

I am continually amazed at the numbers of papers submitted to our journal that are completely outside of our scope. An instant rejection ensues, but I regret the lost productivity that such submissions cost authors and journal staff. To gauge whether a paper will have any chance of actually getting into peer review at a highly competitive international journal, authors must try to honestly assess the novelty and impact of their work and whether or not it provides new insights to the scientific community. CYNTHIA E. DUNBAR, MD, Editor-in-Chief, *Blood*, Journal of the American Society of Hematology

It seems obvious that choosing the right journal for a particular paper is absolutely key to getting your work published, and yet making this critical selection is often where authors make mistakes. In this chapter, we look at how to go about making that choice in a way that maximizes your chance of getting into peer review and getting published.

THE AUTHOR AS MATCHMAKER

Finding a good pairing between a specific manuscript and a specific journal puts the author into the role of matchmaker. The activities of finding a compatible mate in life and that of finding a match between a journal and a specific paper are based on similar principles. Despite the fact that no two people are "exactly" perfect for each other, successful marriages do happen, usually because there is both a strong underlying compatibility and because both individuals' primary needs are met. Similarly, for any one particular piece of research there are probably several journals that are possible "mates," journals in which the paper could be published. Some publications will be a better fit than others, and initially it is the author's responsibility to act as matchmaker and objectively assess both the manuscript and each potential journal to decide which match will best allow everyone's needs to be met, particularly the Editor's.

All the important components to consider in choosing a journal to submit to revolve around the traditional model of effective communication: who is your audience, what is your message, and why is it important? The more clearly you understand your audience, your message, and your purpose in publishing, the better the match you will make when you target a journal for your manuscript. In addition, having a particular audience in mind—not only when you are selecting a journal for submission, but before, when you begin to write and even when you begin your research—allows you to structure your information into a coherent form to get a specific argument across about what your data mean.

As you are finalizing your manuscript and considering journals for submission, you should be asking yourself:

- Does the journal publish the type of paper I'm writing (e.g., research paper, method paper, data paper, review paper)?
- Is this paper really at the right level in terms of innovation and importance?
- Does the journal cover the particular topic I'm writing about?
- Does the audience I want to reach read the journal I'm considering?
- How prestigious is the journal?
- What is the turnaround time for articles in that journal? How important is it for this paper to appear quickly?
- Is there a cost to publishing in the journal? Do I have the funds available?

Most importantly:

- Does my paper fit the aims, scope, or mission of the journal I'm considering?
- Have I thoroughly reviewed the instructions for authors? Can I meet all the requirements stated there?

Most of the answers to these questions can be found in the instructions to authors and scope statements of each journal, and each journal's requirements will be notably different. You need to carefully and thoroughly study this guiding information for each journal that you are considering. You should be sure to get several back issues of the journal and review them carefully, not only to study the style and requirements, but also to get a good idea of what topics have been covered recently. At the same time, you also need to be clear about your own goals and constraints because these parameters (including, for example, the status of a journal, its typical speed from acceptance to publication, and the costs of publishing) can bear directly on which journals may or may not be a good match for the particular paper you are writing.

STUDY THE SCOPE OF EACH JOURNAL

Somewhere on the website, or among the pages of the print copies of the publication (if it includes a print issue), most journals have some sort of statement about the range of topics they cover and types of papers they publish. Some journals call this their "scope," others call it the journal's "aim," or even their "mission." It may be called something else entirely. Regardless of labels, you need to look for language that explicitly states what the journal publishes and why, as well as statements about the primary readers and the main purpose of the journal, as this information can be extremely helpful in understanding what kinds of papers the Editor will consider. A sentence that reads, "The journal is specifically interested in research that has policy implications" is a clear indication that somewhere in your abstract, discussion, or conclusion sections you should include a paragraph or two about the policy implications of your findings. If the main implications of your research do not match the stated requirements of the journal, do not send the paper to that journal. However, if they do match, you should make this clear to the readers, the first and most important being the Editor. If the science has been done properly, you'll have the best chance of getting your paper reviewed if your topic matches the scope and falls within the range of interests of the journal.

Editorial collaboration: the same publisher

Many publishers produce a number of different journals, sometimes all covering one subject (e.g., Ecological Society of America) or else one general journal, covering a broad range of topics, together with a series of specialty journals (e.g., *Nature* and its family of topic-focused review journals). If there is nothing basically wrong with your paper except that, for instance, you aimed a little high in terms of importance or novelty, or you chose a journal that is more focused on applied science than your paper really warrants, the Editor may reject the paper but may either strongly encourage you to resubmit to one of its sister publications or may offer to pass it directly to the other, more suitable journal, via a shared online submission system. In the latter case, you will first be asked whether you agree to allow the manuscript to be passed to the alternative publication.

In some but not all cases, even within the same family of journals, the original Editor may ask the new Editor whether he or she is interested in seeing the paper and obviously won't suggest this option to you if the answer is "no." However, if you do receive such a suggestion, you need to be aware that this in no way guarantees that the paper will be accepted or even that it will be sent to peer review. Also bear in mind that the situation described here applies only where the journals are from the *same* publisher. If the Editor simply suggests that you try submitting to another journal or another type of journal, this does not mean that the other Editor has been contacted—it is just a suggestion (see also chapter 10).

Editorial collaboration: different publishers

Some journals, usually but not always smaller or less well-known publications with a lower or no impact factor, may allow you to submit a paper that has been rejected from another journal, together with the anonymous reviews you received from that publication (clearly, you will need to have made the changes listed in the reviewers' reports). This can be a path to rapid publication following rejection, since these journals will not necessarily send the paper out to reviewers of their own. It will be up to you to judge whether the advantages of getting your paper published quickly outweigh the need to publish in a higher-impact-factor journal. You can find out journal policies on this strategy by checking journal websites and instructions to authors.

In the future, the practice of sharing reviews between journals may be formalized and extended within groups of journals that have agreed to form consortia, and there are already experiments looking at possible mechanisms for achieving this type of collaboration. If these experiments are successful, both authors and Editors could benefit in terms of decreased time to publication and reduced duplication of effort during the review process.

FIND AND CAREFULLY READ THE INSTRUCTIONS TO AUTHORS

The first place to look for the instructions to authors is the journal website, although some publications also provide a shortened version in at least one print issue a year. Good instructions to authors will include:

- journal scope statement (although scope, aims, and mission statements may be presented separately)
- descriptions of the various article types published
- limits: total manuscript length (word or page limits), number of citations, number of tables and figures, or other accompanying materials
- format requirements: type sizes, font preferences, file type requirements for figures, instructions on preparation of tables, use of line numbers, what goes on the title page, and other specifics

- style preferences: e.g., use of active or passive voice, particular words or symbols, reference styles (within text and in the references section), and other language requirements
- information required: e.g., author information, abstract and keyword requirements
- rules on the use of acronyms and abbreviations in the body text

One important note here is that we say "good" instructions to authors will contain this information. In reality, not all journals include all this information directly under the heading of "instructions to authors." You may need to look for it elsewhere on the journal website and you may even need to send an inquiry to the editorial office. Not all instructions to authors are well written, complete, or helpful.

Many authors are unaware of, misunderstand, or simply ignore the instructions to authors, and Editors can usually tell if this is the case as soon as they open a file. In some cases, really blatant disregard of the instructions can result in a paper going immediately onto the reject-without-review pile, although more usually the manuscript will just be returned to the author for reformatting and adjustment, causing unnecessary delay. Even if the topic and paper type are appropriate for the journal and the paper looks interesting, an Editor may still choose to reject it if it deviates too far from the journal's formatting and other requirements. At the very least, disregarding the journal's instructions will irritate the Editor and give a bad impression of the author. We cannot stress enough the importance of carefully studying and following instructions to authors, including authorship requirements, guidelines on conflicts of interest, policies on embargos, and other specifics related to clinical trials registration, data deposition, statistical analyses, and other parameters laid out by each journal.

Again, remember that instructions to authors can vary substantially from journal to journal. Every time you send a paper to a journal—whether a first submission, or a submission following rejection by a previous journal—always study the new publication's instructions to authors and change the manuscript accordingly. Although submitting a paper in the wrong format may not be grounds for rejection, except in extreme cases, it will annoy the Editor and should be avoided. In many cases, the manuscript will be returned to you with a request to change it to the required format, which is a waste of everybody's time, including your own. Many authors ask whether it is really necessary to keep changing the manuscript, according to each new journal's specifications: the answer is *yes*, it is absolutely necessary.

PRESUBMISSION INQUIRIES

Some journals allow authors to make a "presubmission inquiry" before formally submitting their manuscript; sometimes, though not always, these journals have specific requirements regarding what you should send. Some require an abstract, while others ask for a synopsis of the paper. Some journals want a cover letter that answers specific questions, while others don't. You may be asked to provide the full introduction to the paper, together with citations and perhaps all the figures. Every journal has different requirements, so find out what these are.

Even when the journal does not specifically encourage presubmission inquiries, some authors contact the editorial office of a journal on an informal basis if they are not sure their paper is a good match for that publication. Editors have different views on this sort of informal inquiry: some see presubmission inquiries as a chance to save themselves and the authors a lot of time and trouble if the paper clearly isn't a good match. Others feel they don't have the time to read or respond to inquiries "outside the system," or don't want to make a judgment based only on an abstract or brief summary. Some Editors are also reluctant to give a positive response to an informal inquiry, knowing that once formally submitted, there is no guarantee that the paper will be sent out to peer review.

If you decide to write to an Editor, be sure to indicate how your paper meets the journal's requirements. Do your best to make your inquiry short and clear, and explain what it is about your paper that is new, interesting, and important to the readers of that journal. The good news is that most journals that do allow presubmission inquiries try to respond to them quickly—often within a few days to a week. In other words, making such an inquiry will not cause you to lose a lot of time and, with luck, you'll get some useful feedback.

Remember though, as mentioned above, if you do get a positive response to a presubmission inquiry, it does not mean that a paper will definitely be accepted for peer review. But whether the response is positive or negative, replies to a presubmission inquiry may provide useful information that you can use to evaluate the potential match between your manuscript and that particular journal.

IMPACT FACTORS

The strengths and weaknesses of the impact factor (IF) system and other metrics of impact are reviewed in detail in chapter 6. However, many authors put too much weight on this metric when they are selecting a journal for submission, so it is worth a brief discussion here.

All researchers would like to get their work published in a prestigious journal with a high IF and, as a result, many authors, particularly international researchers or early-career scientists, regard the IF as one of their most important criteria when selecting a journal for submission. Publishing in a journal with a high IF yields all sorts of positive benefits for authors. First, of course, publishing in a high-impact journal is a status symbol for a researcher, one that can give an advantage in getting tenure or promotion, can help researchers in the quest for funding, and can help boost general recognition in scholarly communities. In some countries, researchers get sizable financial bonuses for publishing in high-impact journals. Getting a paper into a high-impact journal adds to the overall weight of a researcher's publication list and may help in getting future papers published. It is also seen, often incorrectly, as an indicator of the overall quality of a researcher's work (see chapter 6 for a discussion on the incorrect uses of IFs).

However, putting too much weight on the IF when targeting a journal for submission can have a number of disadvantages for authors. First, high-IF-rated journals usually have very high rejection rates—particularly rejection without review. For example, consider the initial reject-without-review rates for 2010 provided by the editorial offices of three well-known, high-impact journals:

- *Nature*—78%
- *Science*—74%
- *Proceedings of the National Academy of Sciences*—51%.

Even more papers are rejected later in the review process. These are not great odds. Of course, other high-IF journals are not quite as bad as this, but a 50–75% rejection rate is not out of the ordinary. You need to think carefully and objectively about the relative importance of your findings and the chances of getting past that first hurdle—reject without review. Is your paper really important enough to stand out from all the competition? Although we encourage all authors to be optimistic, that optimism must be tempered with realism: the goal is to get your work published in the best possible journal, and that may not be one of the highest-ranking journals in your field.

The point is that high-IF journals usually want to publish truly groundbreaking science. You need to consider—as the Editors will—where your paper fits along the IF spectrum.

SPEED TO PUBLICATION

When researching journals, keep in mind the time frame in which you need to publish. For some authors, speed may not be particularly important; in other cases, time to publication will be crucial. For example, if you are hoping for a new job or a promotion, you may be eager to add new titles to your publication list as quickly as possible. Similarly, if you are working in a fast-paced field, you may be concerned about time from submission to print because essentially you are in a race with other researchers to publish your findings first. There may be other reasons to want to publish quickly, such as an upcoming conference or a pending decision by local or state decision makers that you hope to influence.

Publication time has, in fact, become so important to authors and Editors that many journals are changing their internal processes to allow papers to be published more quickly. As recently as ten years ago, getting a paper into print within a few months was considered very rapid publication. Today, it is not uncommon for accepted papers to be published online within a few weeks or even days. Some journals, such as *PLoS One* and *Ecology Letters*, were launched with the specific mission of rapid peer review and fast publication. Many other journals are adding new sections to their websites that allow accepted papers to be viewed online before final editing (papers in press) or before print (pre-press publication). Yet, despite all these developments, some journals remain notoriously slow at getting papers out to their readers.

Many publications post the average time from acceptance to publication on their websites. If speed is an important factor for you, and you cannot find information on turnaround times, you can try writing a quick note to the editorial office to ask. When you submit, you can also include the reasons why timing is important in your cover letter and hope that the Editor will make an effort to expedite the paper through the system.

FINANCES

Money is another constraint that you need to keep on your list of potential barriers to publication. If your financial resources are limited, you need to find out all the possible costs associated with publishing in a particular journal, including page charges, color charges, and fees related to submission or open access (see sidebar 5.1).

SIDEBAR 5.1

Choosing open access for your paper

CATRIONA MACCALLUM
Senior Editor, *PLoS Biology*, and Consulting Editor, *PLoS ONE*

WHY OPEN ACCESS?

The almost exponential growth of open access journals and articles since 2002 means that where you choose to publish your research has consequences you may never have considered. The choice between a subscription journal that restricts access and an open access journal that removes all barriers to access is important to you as an author because it will affect the way your paper and the data associated with it are disseminated. It is also important to understand that free access (i.e., free to read) is not the same as open access. Open access means that your article can be read, downloaded, and, crucially, reused by others without permission, as long as you receive appropriate acknowledgment for the work. It therefore affects how the ideas, figures, data, and analyses in your study can be used in the future as well as how that information can be linked to other relevant literature or incorporated into databases or websites that can be used by other scientists, policy makers, and the public. The growth of sophisticated search engines and freely available analytical tools can enrich science and research, but only if there are no barriers to the access and reuse of literature and data. Choosing to place your article behind a subscription barrier therefore limits the potential impact of your work beyond just restricting the readership.

In the early 2000s, open access publishers such as BioMedCentral (BMC), the Public Library of Science (PLoS), and Hindawi (an academic publisher that supports more than 300 open access journals) set out to show that a one-time charge for the cost of publication (the "publication fee" model) could provide an alternative and sustainable business model to that of subscription or "'toll-based" publishing, while still maintaining rigorous standards of peer review. Open access publishing is now an established part of the publishing landscape. More than 6,300 journals are listed by the Directory of Open Access Journals (http://www.doaj.org), and not all of them charge for publication. PLoS, BMC, and Hindawi are perhaps the best known of the publishers that charge a fee for publication in their journals. Many traditional publishers also provide authors with an open access choice in their subscription journals (e.g., the "hybrid" journals provided by Springer, Wiley-Blackwell, and the US National Acad-

emy of Sciences). Other publishers have launched their own dedicated open access journals. In the past year, notable additions include Nature Publishing Group's Scientific Reports, AIP Advances, SAGE Open, BMJ OPEN, Open Biology *(from the UK Royal Society), and the Ecological Society of America's* Ecosphere.

WHAT TO LOOK OUT FOR WHEN CHOOSING AN OPEN ACCESS OPTION

Open access publishing, despite its rapid growth, is still an evolving business model, so if you choose to make your paper open access you should understand the different options that are available. Not all journals that claim to be open access really are open access and, as with subscription journals, the quality of the end product and the services the publisher provides can vary. The following is a brief guide to what you should look out for.

Is the journal or publisher option genuinely open access?

The Open Access Scholarly Publisher's Association (OASPA, http://www.oaspa.org) recognizes a journal as open access if it provides free, immediate, online access to all original research and allows reuse of the work in any digital medium for any responsible purpose, subject to proper attribution of authorship. OASPA encourages journals to adopt a Creative Commons License (http://creativecommons.org/licenses) as a standard, in particular the most liberal "attribution" license (a CC-BY license is the least restrictive license; it allows users to redistribute, adapt, or build upon your work, even commercially, as long as they give credit to the author).

Be aware that there are different creative commons licenses and not all of them are compatible with open access. PLoS, Hindawi, and BMC all use the CC-BY license for their articles, as do journals such as SAGE Open *and* Ecosphere. *Other journals, such as* Nature's Scientific Reports, *provide you with the option of two creative commons licenses. Both restrict commercial reuse, and one does not permit any other derivative reuse. Under OASPA's current guidelines above, these options are not open access (see also Carroll 2011).*

Check who retains copyright

If your article is published with a creative commons, open access license, then you should always hold the copyright. The license sets out the conditions under which your article can be used by others, whereas the copyright holder determines the terms of the license and can waive any restrictions (e.g., commercial reuse).

The practice of copyright transfer used by subscription journals is often presented as being in the interests of the author, for example to protect the integrity of their work (see sidebars 8.1 on Why Transfer Copyright? and 8.2 on The Permissions Process), when in fact the subscription business model requires that the publisher restrict the rights of the user. Open access, by contrast, releases your work from all restrictions on reuse (e.g., for photocopying or for course packs), although your work will still be protected from misuse (such as plagiarism) by any reputable publisher, whether open access or subscription-based.

Sometimes, it is hard to find the license or copyright holder on a published paper. Most open access journals put the identity of the copyright holder next to the license (as below) while others position the license and copyright holders separately (e.g., the copyright holder may be found at the bottom of the page in the pdf or online version). The following is an example of what an open access attribution license looks like when the author is also the copyright holder. This is taken from an article in the journal PLoS Biology:

> Husby A, Visser ME, Kruuk LEB (2011) Speeding Up Microevolution: The Effects of Increasing Temperature on Selection and Genetic Variance in a Wild Bird Population. *PLoS Biol* 9(2): e1000585. doi:10.1371/journal.pbio.1000585
> **Copyright:** © 2011 Husby *et al.* This is an open-access article distributed under the terms of the Creative Commons Attribution License, which permits unrestricted use, distribution, and reproduction in any medium, provided the original author and source are credited.

Check that your published article is deposited in a publicly available repository

All open access articles should be deposited in a publicly available repository at the time of publication. This practice complies with the policies of all major funding agencies, including the US National Institutes of Health (NIH) in the US, the Wellcome Trust and the Research Councils in the UK, and the Deutsche Forschungsgemeinschaft in Germany, all of which request or require deposition of the published articles based on research they have funded into publicly available databases. This also ensures that your article is archived stably and in perpetuity (e.g., by being "mirrored" at appropriate sites) and is not held only by the publisher.

Open access publishers, such as BMC and PLoS, or traditional publishers such as Springer that provide some open access options and journals, will ensure that your article is submitted to PubMedCentral (http://www.ncbi.nlm.nih.gov/pmc, funded and maintained by the NIH). Because your article is open access, you can also deposit the final version in a repository of your choice, such as one linked to your academic institution.

Check the services you will receive from your publisher

As an author, you need to choose the most appropriate venue for your paper and, like subscription journals, open access journals differ in the services they provide. Peer review at some journals, such as PLoS ONE, Scientific Reports, *or* SAGE Open *aims to assess the technical rigor of the work submitted and to ensure that the conclusions of a paper are supported by the data; however, Editors are not asked to assess the importance of the work. Such journals aim to provide a rigorous and fast evaluation process. At other journals, the peer review process aims to evaluate the relative importance of the work for the field, as well as the technical rigor. Because rejection rates are higher at such journals, publication fees also tend to be higher.*

You should also check what services the publisher provides once the paper has been published, as this may affect how your article is used or evaluated by others. Because technology and web-based tools are changing rapidly, it is likely that publishers—and especially open access publishers—will increasingly provide different postpublication services. For example, PLoS currently provides a range of metrics about your article, which allow readers to track its usage and citation (http://article-level-metrics.plos.org). The purpose of such article-level metrics is to encourage readers to judge articles on their individual merits, rather than as part of the journal in which they happen to be published (see sidebar 3.2 on using DOIs to foster article-level citation metrics).

What is the policy of your funding agency in relation to open access?

Most funding organizations now require their grantees to make their papers, and increasingly the data resulting from their research, publicly available within a predefined time (note that policies about data and papers are often different). Policies are changing rapidly and it is therefore important that you check what your funding agency requires before you submit your paper to any journal. It is your responsibility as the grantee to comply with the demands of your funding agency. Generally, agencies will stipulate how long after publication an article must be made freely available. For example, the UK's Wellcome Trust requires that authors make their articles freely available within six months of publication, but encourages immediate access by providing additional funds to cover publication fees in open access journals. Others, such as the NIH, currently stipulate free access after twelve months. The most effective way to comply with such requirements is to publish your article in an open access journal or to choose an open access option where available.

If the publisher charges a fee to make your paper open access, and your funder provides you with the money to pay for the cost of publication, tell the publisher that funds are available when you submit the paper. Most open access journals will also have a waiver policy for those without grants or other means to pay for publication. At reputable open access journals, Editors are not privy to this information—the editorial decision about the suitability of a paper for the journal should always remain separate from an author's ability to pay.

QUICK LINKS TO OA PUBLISHERS/JOURNALS MENTIONED

BioMedCentral: http://www.biomedcentral.com
Hindawi: http://www.hindawi.com
Public Library of Science: http://www.plos.org
AIP Advances: http://aipadvances.aip.org
BMJ OPEN: http://bmjopen.bmj.com
Ecosphere: http://www.esajournals.org/loi/ecsp
Open Biology: http://royalsocietypublishing.org/site/openbiology
SAGE Open: http://www.sagepub.com/sageopen/landing.sp
Scientific Reports: http://www.nature.com/srep/marketing/index.html

FURTHER READING

Boyle J. 2008. *The Public Domain: Enclosing the Commons of the Mind*. 1st ed. New Haven: Yale University Press.

Lessig L. 2005. *Free Culture: The Nature and Future of Creativity*. NY: Penguin (non-classics).

Suber P. (n.d.) *Open Access News*. Available: http://www.earlham.edu/~peters/fos/fosblog.html. Accessed 27 May, 2011.

Willinsky J. 2006. Access principle: the case for open access to research and scholarship. Cambridge: MIT Press.

Journals published by societies will often charge lower fees to members, or provide publication grants, so it is worth doing the math to figure out whether it is cheaper to pay the membership fee and get reduced publication fees or not. Some journals offer discounts or waive the fees altogether for authors from developing countries, for students, or sometimes even for early career researchers. If publication costs are beyond your budget, you can approach the Editor before you submit, explain your situation, and ask whether the fees could be waived or reduced if the paper is accepted. (Make it clear that you understand that an agreement to reduce or dispense with the fee is not an agreement to publish the paper.) If the answer is "no," then you probably should take that journal off your list of potential targets.

OPEN ACCESS

Consideration of cost is by no means the only reason to think about publishing in an open access (OA) journal. As discussed in detail in sidebar 5.1, publishing in an OA journal is an important option to consider because these differ significantly from traditional journals in the ways they allow readers to access and use their content. Your primary concern as an author is to get your work published, whether in a subscription-based or an open access journal. At the same time, you need to consider who will be able to access your paper and how they will be able to use the information. As the venues for scientific publishing evolve, and with them the restrictions on use and reuse of information, researchers need to fully understand the pros and cons of publishing in OA journals so they can make informed decisions about whether this kind of journal is a good match for their paper and their goals for publication.

OTHER CONSIDERATIONS

A number of possible problems related to authorship requirements may prevent a paper from going forward to peer review. First, some journals have restrictions on the number of authors allowed for each paper—if the number of coauthors exceeds this limit, you will need to write to the Editor and ask whether an exception can be made.

Many journals are also asking authors to acknowledge that they have followed the journal's requirements concerning potential conflicts of interest. Such requests have become much more common in journal publishing over the past several years, in part because instances of scientific misconduct are increasing and have serious implications, particularly in the area of medicine.

Some journals, particularly those published by professional societies, have developed their own policies that define what constitutes a conflict of interest. These publications often specify how potential conflicts of interest must be disclosed. Others defer to professional standards set out by global organizations, such as the International Committee of Medical Journal Editors (ICMJE) or the Council of Science Editors (CSE). For more information on the important issues related to ethics in publishing, see chapter 11. Remember that the author or authors of a paper are responsible for ensuring that all the conflict of interest and disclosure requirements of a journal are met when a paper is submitted. If you cannot get all your coauthors to fulfill the terms related to conflict of interest and ethical conduct, again, you should probably contact the Editor and seek guidance before you submit.

SUPPLEMENTARY DATA AND MATERIAL

In some fields, publishers are increasingly requiring researchers to make their data or their source materials (for example cell lines, DNA, or antibodies) freely available to other researchers in storage facilities or in data repositories. At the very least, authors may be required to register their data in a database, detailing where the original data can be found. These efforts may require institutional support and authorization, so before submitting to a journal with supplementary data requirements, make sure that you have the technical support and institutional permission to fulfill them. (Also see sidebar 3.3 on DataCite.)

THE BOTTOM LINE

The important message here is that every journal is different and what each Editor is looking for is different. Because of this reality, you should never send exactly the same paper to different journals. You need to carefully study all the possibilities for submission, even while you are still doing your research. After you select the journal that you think is the best match for a particular paper, you should study several recent issues of the journal, review its scope, instructions to authors, and then prepare your manuscript and supplementary materials to meet its requirements.

The only bad result in scientific publishing is being rejected without review, as this outcome usually provides little or no guidance on how to do better with the next journal you try. If your paper is sent out to review and you receive feedback, even if your paper is rejected, the comments you get should help you to make your paper better and therefore help you to chart another, and one hopes more successful, path to publication.

CHAPTER SIX

Understanding impact factors

I first mentioned the idea of an impact factor in 1955. At that time it did not occur to me that impact would one day become a subject of widespread controversy. Like nuclear energy, the impact factor has become a mixed blessing.

EUGENE GARFIELD (2000)

If you are pursuing a career in science, you should know what an impact factor (IF) is and what it really means. Every author of a scientific article destined for a peer-reviewed journal needs to understand where these numbers come from, how they are calculated, what they can tell us, and even more importantly, what they cannot tell us. The truth is that many people are using impact factors in ways that their inventor never intended, to answer questions that this metric really cannot answer.

A SHORT HISTORY OF IFs

To understand IFs, it helps to know a bit about why they were originally developed. In 1960, information scientist Eugene Garfield, founder of the Institute for Scientific Information (ISI), launched the *Science Citation Index* (*SCI*), a database that allowed researchers to trace patterns of citations to important scientific papers—in other words, to see who was citing these papers and in which journals the citations were appearing. Garfield and his colleague, Irving H. Scher, needed to decide which of the many available scientific journals to include in the *SCI*. They needed a way to judge which scientific journals were the "best" or most important, and what being "best" actually meant. The answer, they realized, was that the best journals were those publishing the "best" or most important science. The clearest measure of importance of a piece of scientific information is based on how much and how often it influences the work of other scientists—in other words, its *impact* on other people's ideas and research. As all readers of this book should know, if a scientist uses someone else's scientific findings or ideas in their own work, they must cite that scientist when they come to publish their own findings or ideas. So, reasoned Garfield, if one counted how many times research published in Journal X was cited in other people's published papers over a set period of time, and compared that with how many times research published in Journal Y was cited in other people's publications over the same period, one would get some idea of which of these two journals was publishing the more influential science; therefore, one could rank them in order of importance.

In addition to numbers of citations, three factors need to be taken into account when calculating a journal's impact: (1) the number of papers published in each journal, (2) the differences between subject areas, and (3) the time frame.

Number of papers

Let us assume that two journals, X and Y, both have twelve issues a year, and that they both publish papers on marine biology. If X publishes thirty papers in each issue and Y only publishes ten, then X will have a distinct advantage, since it has many more papers that could be cited than its rival, Journal Y. Garfield solved this problem by simply dividing X's citations by 360 and Y's by 120 (that is, by the number of papers each publishes in a year), to level the playing field.

Subject matter differences

Some scientific areas progress extremely rapidly and involve many more scientists, publishing many more papers than some less "vigorous," more slow-moving topics. You could not, for instance, make a fair comparison between an immunology journal, whose contents are in danger of being superseded by new discoveries before the ink has completely dried on the current issue, and an ecology journal, where the experiments being described can take many years to complete and where findings will likely still be valid and relevant many years later. The IF system deals with these disciplinary disparities by grouping journals covering similar topics together; journals in different subject areas are kept on separate lists and are compared only with others on the same list.

Time frame

To make the new ranking system work, there also needed to be a set unit of time within which to count citations to papers within each journal. Two years was chosen because this time frame would show any rapid developments in terms of citation of a paper, even though most papers actually take five years or more to collect their maximum number of citations. Nevertheless, it was felt that two years would provide a more responsive measure, and would be better able to track any changes in the journal, with respect to the other journals in the same category.

THE IF CALCULATION

With all the necessary elements in place, we can now look at the actual impact factor calculation. Within a group of journals, for example on cell biology, the system looks at a specific year, say 2009, and counts the number of citations to papers published in Journal X in the previous two years and divides them by the total number of papers published in X in those two years (as described above). The resulting number is the "impact factor" of that publication. If the same process is repeated for all the journals in the cell biology category, then these can all be compared. The greater the impact factor, the more influence a particular journal's papers are having on the work of other researchers.

Here is an example of the impact factor calculation for Journal X in 2010:

citations in 2010 to papers published in X in 2009 = 168
citations in 2010 to papers published in X in 2008 = 293
TOTAL = 461
citable papers published in X in 2009 = 54
citable papers published in X in 2008 = 54
TOTAL = 108
Journal X's 2010 impact factor: 461/108 = 4.268

The top impact factors for journals in different subject areas tend to differ widely. The highest impact factor recorded to date has been 87.925 for a cancer journal, *CA-A Cancer Journal for Clinicians*. However, this is exceptional; the next highest impact factor is around 50 and belongs to an immunology title; the top medical journals tend to score around 20 to 25, while *Science* and *Nature* jockey for position in the 30s. By comparison, the top ecology journal has an IF of about 14 and there is quite a large gap before the next title on the list. Clearly, it is important to keep journals segregated into their own subject matter categories.

Impact factors for each year are published in June of the following year. In other words, 2010 impact factors are published in June 2011. The reason for the delay is that all the citations appearing right up to the end of 2010 have to be counted and the various calculations made.

IMPACT FACTORS TODAY

Today, the *SCI* is published by a multinational corporation, Thomson Reuters. Every June, they publish the *Journal Citation Reports* (*JCR*), which lists the new impact factors for the science and social sciences journals in their Web of Science, a huge index and database that provides access to many of their other products. The *JCR* also displays a number of other metrics, some aimed at judging the relative influence of the "average" article in each journal and others looking at the way each journal is integrated into the other journals in its category through the network of citations (figure 6.1).

In 2009, Thomson Reuters included 9,162 journals in the *JCR* database. New ones are added each year, and others are discarded if they cease publication or no longer conform to Thomson Reuters' criteria. About 2,000 new or refurbished journals are sent to Thomson Reuters every year in the hope that they will be accepted into the indexed lists, but only about 10% are chosen. If a new journal is among the rejected 90%, then it will not be given an impact factor, which is a major handicap for a young journal just starting out to make its fortune.

If a journal is accepted, it will be assigned to one or more subject matter categories (e.g., a journal may be listed in both the "biochemistry and molecular biology" category and the "cell biology" category.) and then the counting begins. All the reference sections in every peer-reviewed paper in all 9,162 journals are scanned and the citations to each journal are recorded. Once the year ends, figures are checked and rechecked, and in June the new lists are published, giving the impact factor and subject ranking for each journal for the previous year.

WHAT IMPACT FACTORS CANNOT TELL US

Remember that impact factors were invented to provide a measure of the relative importance of different journals in a field—they reflect the average number of citations to papers in a particular journal. They were never intended to be used to make judgments on a particular paper or an individual author. How could they? If you pick out any one paper in a journal, how do you know whether it was cited once, ten times, or fifty times? (Actually, a certain amount of research using Google Scholar or free software called Publish or Perish [Harzing 2011] can provide a rough idea of how many times a paper is cited.) But you still cannot tell, just by looking at that paper, whether it will have a major impact on the thinking of other scientists or not.

Gargouri *et al.* (2010) were looking at the effects on citation numbers of

WELCOME | HELP

2009 JCR Science Edition

Journal Summary List

Journal Title Changes

Journals from: subject categories CELL BIOLOGY [VIEW CATEGORY SUMMARY LIST]

Sorted by: Impact Factor [SORT AGAIN]

Journals 1 - 20 (of 162) [1 | 2 | 3 | 4 | 5 | 6 | 7 | 8 | 9] **Page 1 of 9**

MARK ALL | UPDATE MARKED LIST

Ranking is based on your journal and sort selections.

Mark	Rank	Abbreviated Journal Title (linked to journal information)	ISSN	JCR Data						Eigenfactor™ Metrics	
				Total Cites	Impact Factor	5-Year Impact Factor	Immediacy Index	Articles	Cited Half-life	Eigenfactor™ Score	Article Influence™ Score
☐	1	NAT REV MOL CELL BIO	1471-0072	24057	42.198	38.260	6.307	75	4.4	0.19024	21.973
☐	2	CELL	0092-8674	153972	31.152	32.628	6.825	359	8.7	0.69859	20.124
☐	3	NAT MED	1078-8956	49928	27.136	27.991	5.143	154	6.6	0.19383	12.254
☐	4	CELL STEM CELL	1934-5909	4007	23.563	23.563	6.724	98	1.7	0.04396	14.875
☐	5	ANNU REV CELL DEV BI	1081-0706	8328	19.571	25.533	0.704	27	7.0	0.04372	15.561
☐	6	NAT CELL BIOL	1465-7392	25557	19.527	19.062	4.144	167	5.4	0.17676	11.962
☐	7	CELL METAB	1550-4131	6462	17.350	19.021	2.844	90	2.9	0.06218	9.165
☐	8	MOL CELL	1097-2765	38987	14.608	13.929	2.760	296	5.3	0.30020	9.397
☐	9	CURR OPIN CELL BIOL	0955-0674	13559	14.153	13.634	2.321	109	6.2	0.07645	7.917
☐	10	DEV CELL	1534-5807	14785	13.363	14.058	2.980	147	4.6	0.13558	9.254
☐	11	NAT STRUCT MOL BIOL	1545-9985	19458	12.273	11.670	3.561	180	6.0	0.13241	8.095
☐	12	TRENDS CELL BIOL	0962-8924	10222	12.115	12.642	1.795	78	5.6	0.06093	6.904
☐	13	GENE DEV	0890-9369	54044	12.075	14.198	2.338	269	7.6	0.26719	8.965
☐	14	TRENDS MOL MED	1471-4914	4780	11.049	8.099	1.517	60	4.3	0.02715	3.229
☐	15	J CELL BIOL	0021-9525	71752	9.575	10.121	2.134	313	9.7	0.23124	5.748
☐	16	CURR OPIN STRUC BIOL	0959-440X	9606	9.344	10.015	1.596	94	6.2	0.04925	5.224
☐	17	PLANT CELL	1040-4651	31626	9.293	10.679	1.485	264	6.7	0.11980	4.532
☐	18	EMBO J	0261-4189	74782	8.993	9.395	2.324	321	9.1	0.24818	5.552
☐	19	CURR OPIN GENET DEV	0959-437X	7598	8.987	8.679	0.864	81	6.1	0.04148	4.701
☐	20	CELL DEATH DIFFER	1350-9047	11655	8.240	8.387	2.701	157	4.1	0.06533	3.519

FIGURE 6.1 Various metrics for the top twenty journals in the cell biology category from the 2009 *Journal Citation Reports—Science Edition*, a Thomson Reuters product.

Total Cites: total number of citations in 2011 to papers published over the previous two years.

Impact Factor: See definition and calculation in the main text.

5-year Impact Factor: numbers of citations in 2009 to papers published over the previous five years instead of the usual two years (measures the journal's influence over a longer period).

Immediacy Index: number of citations in 2009 to articles published in 2009 divided by the number of articles published in 2009.

Articles: The total number of citable articles published in the journal in 2007 and 2008.

Cited Half-life: the number of years, going back from 2011, that account for 50% of the total citations received in 2009.

Eigenfactor Metrics: measures the journal's influence by looking at how often it is cited by other influential journals.

open access versus non-open-access publication, but in doing this research, they generated some interesting data on citation rates. Between 2002 and 2006, they collected citation data on 27,197 papers in 1,984 journals listed in the *JCR* and found that:

23% received 0 citations
51% received 1–5 citations
12% received 6–10 citations
8% received 11–20 citations
6% received 21+ citations

These numbers clearly show that many papers receive few if any citations; in other words, the appearance of a paper in a particular journal provides no evidence about the actual importance or future impact of that paper or or its authors.

Employers choosing between candidates for a job opening often take into account the impact factors of the journals in which the applicants have published. The arguments against this practice are the same as above. You cannot tell from the impact factor how influential (or not) a particular set of papers from a certain author were. Happily, some employers are beginning to change this practice. If you do an Internet search, you will find a number of articles and reports on this subject across the fields of medicine and the life and social sciences. The message is slowly spreading that IFs should not be used in this way (for example, see Marder, Kettenmann, and Grillner 2010) and Monastersky (2011).

Various groups of people use IFs for different purposes; some use them correctly, others do not (see table 6.1).

USES, ABUSES, AND DRAWBACKS OF THE IF SYSTEM

Of course, all Editors are keen to improve the IF of their journal, and there are some legitimate ways to do this. When asked what the best method is for improving a journal's IF, Eugene Garfield, developer of the system, usually replies, "publish the best possible science," and of course that is every Editor's aim. Sadly, however, some Editors have been known to try unethical strategies to increase the IF of their publication. For instance, authors might receive decision letters from an Editor, strongly encouraging them to cite the journal they have submitted to more frequently in the paper. Or they might be asked to include fewer citations to a rival journal. If you ever receive requests like this, our advice is to withdraw your paper and submit it elsewhere.

TABLE 6.1. Who uses impact factors (IFs)?

User	*Usage*	*Notes*
Authors of scientific papers	To choose which journal to publish in	This should only be one among many factors taken into account—submission to a high-IF journal that is not a good match for the paper is a waste of time.
	To help decide which journals to read	Again, this should be only one of a number of factors.
Employers	To compare candidates for a job opening, promotion, or tenure	This practice is slowly changing as more and more articles are published pointing out the flaws in this strategy, but it is still a widespread practice.
Grant-awarding institutions	To help decide who should be awarded grants	See Employers, above.
Librarians	To identify the most useful journals for their collection To identify journals to cut from their collections	This practice is also changing; increasingly, librarians are studying usage statistics and requests from patrons to make these decisions, although IFs may still be taken into account.
Editors and publishers	To monitor the success of their publication To assess the effectiveness of editorial policies To compare their publication's influence to that of competing journals	This is a legitimate use of IFs; Editors and publishers can track the progress and improvement of their journals and mark changes due to adjustments in journal policies.
Information analysts and bibliometricians	To look at bibliometric and citation trends and patterns	Another legitimate usage, provided the analyst is aware of the limitations and caveats connected with IFs.

Information drawn from Thompson Reuters: http://thomsonreuters.com/products_services/science/free/essays/impact_factor/

The IF system can be used (and abused) in a variety of different ways, and a number of papers have been published over the years describing these issues and calling attention to various alternative metrics of impact that are being developed and used by the scholarly community. Below we list some drawbacks of the IF system and briefly describe some of the best-known alternatives to it.

- Scientists may be tempted to choose hot research topics, in the hope that this will increase their reputation as "high-impact authors." However, just because a subject is not hot doesn't mean it is not important. In the long run, picking research on this basis could have a negative effect on the whole body of scientific research.
- Editors may suffer from the same temptation, publishing only highly citable papers that are more likely to have a positive effect on the journal's IF, and thereby excluding important but not particularly citable or exciting research.
- Only about 25% of the world's journals are covered by Thomson Reuters, and a high percentage of journals that are included in the *JCR* are in English, which excludes large numbers of foreign-language journals; this is hardly surprising, since the international language of science is English. Nevertheless, this can lead to duplication of research effort around the world and underplays the importance of research reported in non-English-language journals.
- Journals covering small, specialized subject areas in which few scientists are involved will have lower citation rates, giving the impression that these topics are unimportant, which may be far from true.
- Journals that are published in languages other than English, even when included in the *JCR*, may attract fewer citations because fewer scientists around the world will be able to understand and cite them.
- Authors often make mistakes when putting together their reference lists. Although Thomson Reuters has a number of systems in place to try to catch and correct these errors, if a citation includes the wrong year, it may not be counted.
- Dare we say it? Thomson Reuters can make mistakes as well.
- For some of the more slow-moving subjects, a three- or four-year window would reflect citation patterns better.
- Most important of all, IFs can only measure, on average, how influential the papers in a particular journal are on the work of other scientists, but this says nothing about how useful a particular piece of research is in solving real-world problems.

OTHER METRICS

Given all these issues, it is not surprising that various researchers have tried to come up with better metrics to judge the comparative importance of journals, of individual authors, and of the papers they publish.

Probably the best known of these newer metrics is the *h* index, developed by physicist Jorge E. Hirsch (2005), which considers the impact of individual researchers by looking at their publication record, or in other words, their "scientific productivity." The *h* index is defined as the number of articles published by an individual that have attracted citation numbers that are equal to or slightly higher than the number of papers they have published. So, if an author has published fifteen papers that attracted fifteen or more citations, that author's *h*-index is 15. The *h* index therefore looks at both productivity in terms of numbers of papers and impact in terms of numbers of citations to those papers. It is even possible to aggregate scores for all the authors in a journal or all the researchers in an institution and so calculate an *h* score for that journal or institution. For more details on this method, see Hirsch (2005).

Another metric, the eigenfactor, ranks and evaluates the importance of the journals in the Web of Science index and also looks at the influence of individual articles within five years after publication. A free listing of eigenfactors is available at www.eigenfactor.org/index.php.

Finally, the SCImago Journal & Country Rank looks at the rankings and visibility of journals within different subject areas as well as scientific indicators in different countries; rather than being based on the Thomson Reuters database, the journals ranked by this method are all in Scopus, an abstract and citation database published by Elsevier B.V. (SCImago 2007).

FURTHER INFORMATION

If you want to know more about the development and use of IFs than we have had room for here, there are plenty of materials you can refer to, starting with their creator's perspective (Garfield 2006).

The Thomson Reuters website carries information on impact factors: http://thomsonreuters.com/products_services/science/free/essays/impact_factor/. For a history of citation indexing, see http://thomsonreuters.com/products_services/science/free/essays/history_of_citation_indexing/.

For a basic overview of some of the other measures of the influence of journals, scientists, and publications, go to the HealthLinks website at http://healthlinks.washington.edu/howto/impactfactors.html

THE BOTTOM LINE

Authors often place great importance on the impact factors of journals when deciding where to send their paper. Impact factors are also factored into various other important choices, such as which applicant gets a job, who gets tenure or grant money, and so on. It is therefore important to understand how these numbers are calculated, what they mean, and the correct way to interpret and use them.

CHAPTER SEVEN

How to write a cover letter

We are well aware of the word limit at *Frontiers* and we did our best to stick as closely to this as possible, but we feel we can only do the subject justice by writing a somewhat longer paper. We hope you will give us some leeway . . .

Unsuccessful *Frontiers* author

You might imagine that the Editor of the journal to which you have sent your manuscript pores over it for a considerable length of time before making that first, vital decision about whether or not to send your paper to peer review. The truth is, however, that in many cases Editors make that initial determination quite quickly. They will check the basics for each incoming submission, such as whether the topic falls within the journal's area of interest, whether it is the right type of paper for the publication, and whether the manuscript conforms to the criteria specified in the instructions to authors. If a paper has too many problems, the Editor may send out a reject-without-review notice and move on to the next manuscript without delay.

However, if this initial "quality check" reveals no obvious problems, then the Editor still has to decide whether the scientific content merits further attention. Your first goal as an author is to convince that first Editor that your paper should survive this initial cut, and this is where a good cover letter can help swing the decision in your favor. In this chapter we explain how providing a well-crafted cover letter to accompany your manuscript can help an Editor understand the importance and relevance of your paper to their journal, so that they are more likely to send it out to peer review.

THE COVER LETTER AS A FIRST INTRODUCTION

Going back to the metaphor of author as matchmaker, as described in chapter 5, you can think of your cover letter as the first introduction between the Editor and your paper. Everyone knows how important introductions are in everyday life. Your first impression of someone, and the initial information you receive about them, can greatly influence your attitude towards them and whether you want to get to know them better or not. In the same way, if you can catch the Editor's interest right from the start with a strong cover letter, you may be able to persuade that Editor to try to get to know your paper better. In addition, the cover letter provides you with a rare opportunity to communicate directly with the Editor, to explain the importance of your work, and to lay out the reasons why the paper is a good match for the journal.

Another consideration in the Editor's mind when assessing a new submission is the scarcity of a very precious resource: good peer reviewers and the limited time they are able to spend evaluating manuscripts and writing reports. Editors are therefore extremely reluctant to waste the time of their peer reviewers by sending them papers that don't seem very interesting. This means that you only have a narrow window of opportunity to attract the Editor's attention and prevent your paper from being consigned to the reject-without-review pile. A cover letter can help you make the most of that opportunity.

It is true that some Editors don't read cover letters at all, but others do read them, so you have nothing to lose and a lot to gain by writing a clear, short, informative cover letter to accompany your manuscript.

Too short or too long is just as bad as no letter at all

If there is no letter attached to your manuscript, then its fate will depend on the initial Editor's level of expertise in the subject, his or her recognition of the importance and relevance of the information in the paper, how many other papers are under consideration that day, and a number of other, unpredictable variables. If you include only a short note, such as

> Dear Editor,
> I am submitting this paper, entitled [Title], and hope you find it to be of interest and will consider it for publication.
> Sincerely, The Author

this tells the Editor nothing useful, other than the fact that you don't know how to write a good cover letter.

On the other hand, since most Editors don't have much time to spare, they won't want to spend almost as much time reading the accompanying letter as they would spend on the manuscript itself. They may therefore move straight on to reading the paper. In other words, a very long cover letter will probably not be read, so you should not waste time writing one.

Editors who read cover letters prefer them to succinctly offer useful information that will help them make a decision about whether to send the manuscript out for peer review. There are a number of possibilities regarding what that useful information might be, and only a small number of the possibilities listed below will be relevant to your paper. Your task is to choose which two or three of these points will best describe the most significant aspects of your paper and then to build your cover letter around those.

A good cover letter should not exceed one page and should consist of no more than four or five paragraphs. One and a half pages is the absolute limit, and this is acceptable only if all the names and addresses of the authors take up a lot of room. Don't try to cheat by using a smaller font size to squeeze more information onto the page. Remember, the finite quantity in this equation is the Editor's time and patience—once either of these runs out, you have lost your opportunity to make a good impression and to explain why your paper would be a valuable contribution to this particular journal.

The following basic information should always be included somewhere within your initial submission documents:

- the title of the paper
- the names and addresses of all the authors
- the contact information of the corresponding author
- the correct journal name.

Don't be surprised by the last bullet point on this list. Editors quite often see submissions in which the wrong journal name is mentioned in the cover letter or on the first page of the manuscript. This unfortunate mistake can happen in one of two ways. First, the authors think they know the journal name and don't bother to check. Thus, *Frontiers in Ecology and the Environment* incorrectly becomes *Frontiers in Ecology and Evolution* or the journal *Blood* becomes the *Journal of Blood*. This kind of error tends to annoy Editors and immediately gives the impression that the author is careless or lazy. If the author makes an error in something as simple as the journal title, the Editor can't help but wonder whether that carelessness extends to the contents of the paper.

The second way in which incorrect journal titles get into submission documents is when another journal has already rejected the manuscript. Without thinking, the corresponding author copies and pastes the uncorrected cover letter into the online submission system of the next journal on the list. Again, an error in a journal title will not impress the new Editor. No Editor likes to think that their journal was the author's second choice, even when the first choice was *Science* or *Nature,* or the top-ranked journal in the field. In short, you should always change your cover letter when you submit your paper to another journal, no matter how many times that might be.

Below, you will find a list of possible categories of useful content-based information that could be included in a cover letter. Use only the two or three points that are most relevant to your paper—the ones that you believe will be most helpful in persuading the Editor that your paper should be sent to review. The letter could:

- provide background information on the paper (workshop, interdisciplinary collaboration, or other details that provide a useful context to your work)
- give scientific background information (brief)
- explain the article contents (brief)
- list what is new in the paper
- explain why the new information is important
- explain why the paper is being sent to this journal
- explain why the paper is important *now*

When drafting the letter, you need to take a step back and look at your work, as described in your manuscript, as impartially as possible. You must not exaggerate the importance of the research ("this paper will completely revolutionize the treatment of disease X"), nor should you be too modest or ingratiating ("it would be a great honor if you would consider including my paper in your excellent journal"). Instead, be brief and businesslike, and put forward the two or three points that you think are most likely to attract the Editor's attention.

Below we provide a few excerpts from cover letters that contain some key elements, underlined, that you might consider including in your own letter, where relevant.

Background information on the paper

> <u>This manuscript was the product of a workshop, which took place at</u> the Society's annual meeting in Albuquerque, New Mexico.
>
> <u>The basis for this paper is</u> a white paper, produced by the Working Group on . . .

This kind of information tells the Editor something about the context in which the work was initiated.

Scientific background information

> In this manuscript, we attempt to explain why, although the ecological processes that create treeline patterns across the mountain ranges of North America are mechanistically similar and are linked to climate, actual patterns differ greatly. We discuss what this means in terms of treeline shifts in response to climate change.

Here, the authors are telling the Editor, in two sentences, what the paper is about, with just a tantalizing hint about why this might be topical and timely.

Explanation of article contents

> Charcoal is generated in all biomass burning events and is one of the legacies of forest fire. However, to date, it has received very little scientific attention. In this manuscript, we summarize the existing literature on charcoal deposition, ecological function, and storage in forest ecosystems. We also provide an analysis of how forest management influences charcoal formation and discuss the implications for long-term carbon storage in forest ecosystems of the Rocky Mountain region.

What is new?

> Interaction of light with matter is fundamental to science and technology and has led to the development of lasers and optical fibers which form the backbone of modern communication systems. Significant advances in nanofabrication techniques have now made it possible to fabricate sophisticated optical devices with greater functionality. However, active control of light propagations at subwavelength scales still remains a challenge. In this manuscript, we demonstrate an electro-optic switch based on absorption modulation of light with high contrast ratios and record switching speed. This approach should open up novel avenues in the area of active plasmonics and blasmonics and subwavelength photonics.

Here, in one paragraph, the authors tell the Editor how the research fits in with the existing body of research and what new applications and implications the findings might lead to. Of course, you should not claim that something is new or has never been done before if that is not the case, as that would just signal ignorance of the literature and will have the opposite of the desired effect.

What is important?

> Our findings demonstrate that these techniques are safe to use in all patient populations and have direct implications for healthcare management and reimbursement.

Providing a sense of the importance of the work and how it might affect decision making, and that it may have financial implications, is a useful strategy, and might just catch the Editor's eye.

Why this journal?

> These findings should be of great interest to both surgeons and nursing staff as they have important implications for post-operative patient care. We therefore believe *The European Journal of Surgical Aftercare* would be the ideal forum to highlight this new information.

This text not only tells the Editor that the authors know who reads the journal but also explains why the information will be directly relevant to that readership.

Why now?

> Species A, which has recently been discovered in Lake Mead, is highly invasive, causes enormous damage to underwater pipes and other structures, and is very difficult and expensive to eliminate. The new, more cost-effective management methods we describe here are urgently needed to prevent this invasion from spreading further. We think this work is timely and will be of interest, both to researchers and to resource managers.

Here, the authors indicate that they know who reads the journal, why this information will be of interest, and why it is important to publish this information quickly.

DEAL WITH ETHICAL ISSUES IF RELEVANT

The cover letter is an ideal place to let the Editors know that there are ethical issues (e.g., a conflict of interest) that they should be aware of, related to the study or the manuscript. Again, only include such information if it is relevant to your manuscript. The letter should:

- declare conflicts of interest
- assure the Editor that ethical guidelines have been followed (study the journal's instructions to authors)
- suggest peer reviewers you think could review your paper knowledgeably, and mention those you would prefer the Editor did not approach to review your manuscript.

Declare conflicts of interest

> The only potential conflict of interest is that I am the author of one of the textbooks being discussed in the paper. This authorship does not result in significant monetary gain.

This is a mild conflict of interest that would not be likely to worry an Editor.

> My supervisor is the recipient of a grant from the company that manufactures the drug used in this study.

This is a more serious matter; the level of concern may depend on the findings of the study. The important point is that you have carried out your responsibility in alerting the Editor to this issue, so that any decisions can be made in context.

Whatever conflicts of interest do exist, it is important to explain them honestly at the time of first submission. The Editor is far less likely to be concerned about something that you have openly discussed in your cover letter, as opposed to something that is discovered by accident from a different source at a later date. The latter suggests that you have tried to hide the information, which will immediately arouse suspicion.

Assure the Editor that guidelines have been followed

> All authors have seen and approved the final version of this manuscript.

> The text is under 3500 words in length, as specified in the Instructions to Authors.

The first example is useful, since this is something Editors cannot find out just by looking at the manuscript. They occasionally learn that this important step has not been carried out after the paper has been published, when an angry author telephones to tell them so. This can result in bad feelings and various unfortunate consequences.

Although Editors appreciate it when authors save them some time and effort, the second example is unnecessary, since a couple of clicks of a mouse will provide this information. If your cover letter is very short, it might gain you a slight advantage, but if you have already covered most of a page with useful information, it is probably better to stop there.

Suggest peer reviewers to invite or avoid

> Dr. X, Dr. Y, and Professor Z have all done a considerable amount of research in this area and would be very well qualified to review our paper.

> I was co-author on a paper with Dr. X, but this was eight years ago and we have had little contact since that time.

Some journals actively encourage authors to include suggestions for suitable peer reviewers and are also open to requests not to send the manuscript to certain individuals. Other publications give no indication as to whether they welcome such suggestions. In the latter case, it does no harm to add a short list of names, if you feel confident that you can pick suitably knowledgeable, unbiased individuals, as shown in the first example above. However, you need to be careful not to suggest anyone with whom you have a personal connection (that includes supervisors, close colleagues, significant others). If you have anything more than a tenuous relationship with someone you are suggesting as a peer reviewer, you must declare this in the cover letter or somewhere in the online submission form, as shown in the second example above (see also the discussion of conflicts of interest, earlier, and in chapter 11 on Ethical Issues in Publishing). Researchers specializing in a narrow field of study can be a very small group, and the longer you work in that field, the more difficult it may become to suggest reviewers that you have had little or no connection with. Be honest and open about your relationship with your proposed reviewers and let the Editor decide whether to follow your suggestions or not.

> Dr. X and his research group are working on a similar system to my own and there is a certain amount of competition to be the first to publish the results of this new technique.

> Professor Z has heavily criticized my work in the past and disagrees with my approach.

If you are asking for a particular individual to be excluded, the Editor will find it helpful if you briefly explain the reason for your request, as shown in the first example above. However, you should be careful not to sound petty or accusatory; instead, try to state the facts as tactfully as possible.

THE BOTTOM LINE

Although some Editors ignore or only glance at the cover letters that come with manuscripts, others read these letters carefully. You should write your cover letter assuming that the Editor who first sees your paper is one of those who read them. Your letter should be concise and should include relevant, useful, and important information that could attract the Editor's attention and improve the paper's chances of getting into peer review.

CHAPTER EIGHT

Preparing for manuscript submission, or "What Editors wish you knew"

Dear Editor,
We are excited to be submitting our manuscript to *Journal of Applied Ecology.*

Cover letter to the Editor of *Frontiers in Ecology and the Environment*

When airplane pilots climb into the cockpit of their aircraft and get ready for takeoff, they begin by running through a number of preflight checks to make sure that all the mechanical and electrical parts of the plane are in good working order, that there is fuel in the tank, and so on. Before launching your paper down the runway, you will need to do much the same thing. And just as the pilot cannot afford to be inattentive when making those final preparations, neither can you. (See appendix 3 for a basic list of these "preflight" checks.) You can also use the pages of this and other chapters to create your own list and make sure that everything has a checkmark (✓) next to it before you click the "submit" button (or before you seal your manuscript into its stamped, addressed envelope).

Although every journal is different, some of the advice given here and elsewhere in this book may be common knowledge to experienced authors, and many of the suggestions may seem ridiculously obvious. Nevertheless, Editors see these same mistakes every week, sometimes every day, which can be very irritating—and the last thing you want is an irritated Editor looking at your paper. In some cases, neglecting your editorial preflight checks will cause the paper to crash at an early stage: in other words, it will be rejected without review.

Some of these preparations can be done while you are writing your paper, particularly if you have a specific journal in mind. As your work progresses, keep checking that what you are doing is in line with the journal's requirements and then check everything again at the end, once you think the paper is ready for submission. If your paper has been submitted to and then rejected by a journal, you will need to change the paper to conform to the requirements of the new journal. In this chapter, we also pay particular attention to two areas of manuscript submission where authors often make mistakes: submitting correct copyright and permission information regarding materials (e.g., graphs, photos, tables) that have been published previously and providing figures in the correct formats and at the correct resolution.

BASIC PREPARATIONS

Look at issues of the journal before you submit

As discussed in chapter 5, you should always try to study a couple of issues of the journal you are targeting, even if getting access to back issues is not always so easy. Ideally, you should try to review the most recent issues of the journal so you can assess the types of papers being published. If you can't get access to recent issues through your institutional library or from colleagues, you can always try emailing the publications office (contact details can be found on the journal website) and request a sample issue or at least a few pdfs of recent articles. Explain why you need them: "I would like to submit a paper to your journal, but first need to make sure my paper is a good match." Many, though not all journal offices will respond positively to this kind of appeal. Don't direct your request to the Editor-in-Chief but, where possible, to the secretarial, administrative, or editorial staff. The website will often provide contact details. You can also try writing to authors directly to ask for a pdf of their recent paper from the journal you are interested in. Finally, many journals now make their content open access after six months or a year, so you may be able to get free access to older issues. Check on journal websites to see if and when back issues become freely available in electronic formats.

Instructions to authors: find them, read them, and obey them

This is probably one of the most important pieces of advice in this book, which is why we repeat it here. The instructions to authors, together with scope statements and other information that is normally associated with the instructions, are a vital resource. If you ignore the journal's instructions, or neglect to do your preflight checks, and some of the errors described below creep into your submission, you will give the Editor the impression that you are careless, or that you produced your manuscript in a hurry and didn't check it properly. Again, this is not the sort of impression you want to give.

THE COST OF SUBMITTING A MANUSCRIPT

These days, there are a number of possible financial costs involved in publishing a paper. You need to find out if the particular publication you are interested in has page charges, open access fees, an extra charge for color figures, or any other costs. If so, can you afford to pay these costs? And if not, is there a possibility that the fee will be reduced or waived altogether? Some journals will do this if you let them know that you do not have a grant that includes publication costs or if you are a researcher from a developing country.

If the journal is published by a nonprofit society or association that has individual members, it is possible that members pay lower publication costs than nonmembers when publishing in the society's journals. Some societies also offer page grants to their members. If so, do the math—you may find that it is cheaper to join the society and pay the reduced publication fees than to pay the full cost as a nonmember. Of course, there may be other benefits of membership that make joining the society even more worthwhile. Generally, if a journal is published by a big for-profit publishing house (e.g., Elsevier, Springer Verlag), you are less likely to be required to pay page charges than you would with a society publisher. However, this isn't true of open access journals, since these publications are usually based on an author-pays business model, no matter who the publisher is.

WORD COUNTS AND PAGE BUDGETS

Editors frequently receive manuscripts that are longer than the upper limit specified in the instructions to authors. Usually, this is because the author has failed to look at the instructions and is simply not aware that such limits exist. Occasionally, however, authors submit manuscripts that exceed the specified word or page count because they believe that the information in their paper is so important that the Editor will make an exception and allow the extra length. Unfortunately, the Editor very likely won't, for two reasons. First, journals have fairly strict page budgets (the number of pages per year that the journal is allowed to publish). If your paper takes up considerably more than the normal number of pages, then there will be less room for the rest of the papers in that issue; fewer papers than normal in the journal may be perceived as less value for money by subscribers—a perception Editors try to avoid. Second, if other authors see that the Editor has allowed one paper to be noticeably longer than normal, they will demand to know why they cannot be given the same leeway. You'll always be safe if you keep your paper within or at least close to the limits specified in the instructions to authors.

Depending on the overall size of the journal, there may be a certain amount of flexibility in the permitted length—a few hundred extra words added to a total of several thousand will probably be allowed. However, if the journal specifies a maximum of 4,000 words and your manuscript is 6,000 words, the Editor will likely either reject your paper without review or, at the very least, will send it back to you for shortening prior to peer review.

Unless the instructions to authors say otherwise, the manuscript should always begin with a title page. This should include the title of the paper, the name and contact details (including email, phone, and fax numbers) of the corresponding author, and the names and addresses of all the other authors, as well as the date of submission.

GENERAL FORMATTING AND STYLE

You will need to look for any special instructions on writing style. For instance, unless submitting to a very narrowly focused journal, you should avoid or explain all specialized terminology. There is usually an instruction to write clearly and concisely, but very few authors seem to understand what that really means, so here are a few tips:

- Avoid long, convoluted sentences—reasonably short, clear, straightforward sentences are always appreciated by readers, even when the subject itself is complex.
- Don't be tempted to use long or fancy words when short, everyday ones will do just as well—this will not impress the Editor.
- Use active voice. Unlike days of old, when passive voice was thought to be a necessary characteristic of technical and scientific writing, most Editors now prefer that manuscripts are written in the active voice when possible. Active sentences also usually end up being shorter than sentences written using passive verbs. (See appendix 1 for online writing resources where you can learn more about active and passive voice.)
- Pay attention to grammar and spelling—again, you do not want to give the impression that you are careless and inattentive to detail.

If your English is poor, you should try to get help from someone who speaks and writes very good English. If you do not know anyone who can help you, perhaps your supervisor or other faculty members can suggest someone they know. Ask them if they would be willing to write on your behalf, asking for help. If the English in the manuscript is poor and the Editor has difficulty understanding the text, your paper may be rejected without review or returned to you with a request to improve the English. This is not because of any language bias, but simply because the manuscript is so hard to understand that no accurate judgment can be made about the scientific content. Make every effort to get the English in your paper to as high a standard as possible, to ensure that Editors and reviewers can focus on the science in your manuscript rather than having to struggle with the English. You might consider sending your paper to a commercial language-polishing service. Some journal websites or instructions to authors now provide links to one or more of the many companies that help authors with their written English, for a fee. See sidebar 8.1 for information on how to go about choosing a good language-polishing service.

SIDEBAR 8.1

Choosing a good language-polishing service

MARY ANNE BAYNES
Director of Sales and Marketing, The Charlesworth Group

For many authors who are nonnative English speakers, writing papers in English can be difficult. If you have an English-speaking colleague or friend who can read through your manuscript to make sure it is written correctly and clearly, that's one way to get help. However, not everybody has such a colleague nearby, and not all native speakers of English are necessarily good writers (far from it!).

To get help with your written English, you can also send your manuscript to a professional language-polishing (or language-editing) service. Using a professional service can provide a quick and accurate language review and revision of your manuscript and can usually do this better and more quickly than a colleague. However, you need to choose a service that guarantees high-quality work at a reasonable price.

A good language-polishing service should guarantee that they will not only correct your spelling, punctuation, and grammar, but also that they will review sentence structure, overall paper structure, flow, and consistency—those are the characteristics of language that allow the science to be read and understood clearly. High-quality and reputable language-polishing services should also guarantee that they will not make edits that will change the meaning or the science in your paper and, importantly, not *guarantee that your paper will be published.*

Some publishers provide lists of polishing/editing services on their websites and this is a great place to start. Some publishers actually closely review these companies and know that they do good work. More often, however, publishers state that the services they list are not vetted and that the author is responsible for checking the quality and reliability of the editing service they select. Whether you select a service from a list the publisher provides, or from doing an Internet search for "language-polishing" or "language-editing" services, you should always study the websites carefully to see what they say they will do and what they won't do.

When you are selecting a language-polishing service, ask the following questions:

- ***Exactly what will I be charged for the work?***
  Reputable companies use a number of options for pricing. Some will list their charges clearly on their website. Some charge by the word, while others price according to ranges (e.g., 1,000–5,000 words will cost a certain amount). Other companies will ask that you send your paper in for an estimate and will then quote a price. Costs can vary greatly, so make sure you understand the pricing structure and know what you will be charged before you agree to let them begin work. Don't agree to work before you know the cost.
- ***How long will the edit take? What is the turnaround time? Are weekends and holidays included in the turnaround time?***
  Some companies have preset turnaround times, so you know when your paper will be returned; other companies will provide a turnaround time once they've had a chance to look at your paper. Some companies will charge a premium for a fast turnaround. Make sure you know what the final turnaround time will be and that it will work for your schedule.
- ***What is included when editing? What is done?***
  Make sure you understand what is included with the service before they start the work—what will be done to your paper and what will not be done. Will they check the format of references, or the labels or captions of figures and tables?
- ***Are* all *the Editors native English speakers?***
  If you are trying to publish in English, select a service that employs only native English-speaking Editors.
- ***Do you need American or British English?***
  Make sure the editing service edits in the style you need for the particular journal you are submitting to and that the Editors are able to handle the style you require. American and British English have different rules for grammar, spelling, and punctuation.

- ***Where is the editing done? The US? India? The Philippines?***
  Language-polishing services may use Editors from all over the world. Some of the Editors may be native speakers of English but may not live in a country in which English is the primary language. Knowing the country where your Editors are working will give you a better idea of their language level.
- ***Are the Editors trained to degree level or above in the manuscript subject?***
  The best language-polishing/editing services employ Editors who have degrees in the subject matter of the papers they are editing. You should select a company that can offer Editors who have some academic training in the area you are writing about.
- ***Do the Editors go through a qualification/training process?***
  The best language-polishing/editing companies test and train their Editors to make sure their editing skills are at a very high level. The company website should clearly state that all the Editors are trained and highly qualified. High-quality editing by professionals who are familiar with your subject is what you are paying for!
- ***Is there a guarantee? What if I don't like the result of the service?***
  A good language-polishing company will offer a guarantee of their work.
- ***Is my research secure?***
  Make sure the service has a secure submission system and that the Editors have nondisclosure agreements in place to keep your research safe.
- ***How good is the online submission system?***
  Check out the online submission system of each company that you are considering. Is it easy to use? Is it available around the clock?

Language-polishing services can vary in their offerings, but with a little research and the right questions, you should be able to find a good service that fits your needs and is within your budget.

Appendix 1 includes lists of websites as well as excellent textbooks on science and technical writing. Some of the websites provide tutorials on writing and editing. If you are not confident in your writing skills, whether you are a native or a nonnative speaker of English, exploring these writing resources would be well worth your time.

GENERAL CHECKLIST

Here are a few more things you can do to please the Editor. These are all related to problems that Editors see regularly.

Line numbers

Most Editors and peer reviewers strongly prefer to see line numbers in a manuscript. To do this in Word, go to the "page layout" menu and click on "line numbers." Note, though, that you should always follow the journal's submission guidelines, as some journals differ on this matter.

Latin and other proper names

You need to check that all spelling is correct and consistent throughout the text. For instance, check that all genus and species names are spelled correctly each time they are mentioned. Authors often misspell Latin names or spell them correctly the first time and then wrongly elsewhere in the manuscript. The same goes for chemical names, drug names, and place names. Errors of this kind give a very bad impression.

Numbers

Check that numbers are consistent throughout the manuscript. For instance, if 560 plots (or patients, or plants, or experimental animals) are described in the materials and methods section, make sure this is also the number given in any associated tables and figures, and that the same numbers are discussed in later sections of the paper. Always account for any discrepancies (e.g., 500 patients were enrolled in the study, but 18 failed to fill out the questionnaire correctly and so were not included in the analysis).

Acronyms

You must remember to spell out all acronyms when they are first used in the manuscript, followed by the acronym in parentheses, as follows: Bureau of Land Management (BLM), human papillomavirus (HPV). After that first mention, you can use the acronym alone. When authors use particular acronyms on a regular basis at work, to the point where these become part of their everyday language, they can sometimes forget to explain what the letters stand for. Occasionally, the authors themselves forget what an acronym stands for or, more frequently, what the correct spelling should be (e.g., World Health Organization, with a "z," not an "s," or United Nations Environment Programme, not Environment*al* and with "mme" instead of a single "m"). Remember also that journals differ in their rules on the use of acronyms. Some publications forbid their use in the abstract, preferring that the words be given in full. Even if you are allowed to use acronyms in the abstract, you will still need to give the full term, followed by the acronym, at its first use in the main text. The same process may need to be repeated in the figure captions because the abstract and the main text of a paper, and sometimes the figures, may be seen in isolation by readers.

Presenting citations

Make sure that the citation style is correct for the journal you are submitting to, both within the text and in the references section. Some software packages (e.g., Endnote, ProCite) can help by automatically changing references to follow the correct style of many well-known publications. However, when using this type of software, it is important to check your reference section manually, since these programs can introduce errors.

Authors often ask, "Why do I have to change the reference style every time I send the paper to a new journal? It's a waste of my time—the journal staff can change the style after they have accepted the paper." Now look at it this way: do you really want the Editor to know that your paper has just been rejected by another journal? Or that you didn't see his or her journal as the best possible match for your paper, but instead chose a different publication? Is that going to make a good first impression? Many journals have quite distinctive citation styles, and Editors who recognize that their journal was not the first publication you thought of when you wrote the manuscript are likely to start wondering what it was about the paper that the previous Editor didn't like.

Here are some other extremely common but easily avoidable errors that authors make in reference sections. Before you submit, but after all other changes have been made, check your references carefully. Failing to do so wastes the journal staff's time and annoys the Editor. In particular:

- Make sure that all your citations are correct by checking the details against the original papers; don't simply reproduce the citation that you see in another publication, since it could be wrong.
- Make sure that each DOI (digital object identifier) is correct; it is very easy to make mistakes when typing out these long strings of letters and numbers. Use the DOI resolver on the CrossRef website (http://www.crossref.org/05researchers/58doi_resolver.html) to check that they are correct. If the paper has a CrossMark symbol, check that the paper you are citing has not been amended or even retracted (see sidebar 3.2 on CrossRef and CrossMark).
- Ensure that citations in the main text all appear in the references section and that all the citations in the references section also appear in

the main text, and that spellings of the authors' names and the dates of the citations are the same throughout your manuscript.

- Make sure that all Internet links are correct and functional. Some journals will request a "viewed" date (i.e., the date you last checked that the link was working).

DEALING WITH FIGURES AND TABLES

Editors see a lot of easily avoidable mistakes related to figures and tables. These often have nothing to do with the scientific content but are mostly connected with formatting and submission requirements.

Are all figures present and correct?

Problems often arise because of the many drafts that a paper goes through before it is submitted. For instance, let's say you write a paper that has six figures. Your colleague or supervisor suggests that Figure 4 is unnecessary, so you remove it but forget to also remove all references to it in the text (the mention of a figure or table in the text is sometimes called a call-out). You must be careful to renumber all the subsequent figures and their call-outs so that everything is in the correct order and all call-outs refer to the correct figures.

Another common error arises when paragraphs and sections of text are moved around during all the drafting and redrafting of a manuscript, particularly when the draft is being passed among multiple authors. As a result, figures can get out of order so that, for example, figure 5 is mentioned in the text before figure 4. One of your very last preflight checks should be to make sure that numbering of figures and tables is correct and that everything appears in the right order. The easiest way to do this on a computer screen is to highlight each Figure so that it stands out clearly within the text and you can easily check numbering and order (just remember to remove the highlights before submission). On paper, use a light-colored highlighter pen for the same purpose, but don't send the highlighted copy to the journal.

You should also make sure that all the figures and tables you refer to in the text are provided together with the manuscript and that all the figures or tables you provide are mentioned in the text. The Editor and peer reviewers will not appreciate finding a stray figure that does not appear to be referenced anywhere in the paper, nor will they be pleased to read an interesting description of figure 6 when figure 6 is nowhere to be found.

Are you sending the correct version?

You and your coauthors need to make sure that the figures or tables you submit are the latest versions. Many an Editor has received a frantic phone call from an author, usually just before or just after the paper has gone to press, because an earlier draft of one of the figures was submitted by mistake. Since all authors must see and sign off on the final manuscript and all its associated figures and tables, someone should have spotted that the wrong version had been attached. In fact, authors will often see what they think should be there rather than what actually is there.

Resolution

One section of the instructions to authors that seems to confuse a lot of authors is the one about the required formats for photos and graphic illustrations and, in particular, the specifications on resolution. Journal staff seem to spend more time helping authors with this aspect of the submission process than any other. Explanations about resolution (which can be found all over the Internet), can get rather technical, so this description of the subject is going to be very basic—a brief overview of what authors need to know about submitting photos and graphics with their manuscripts.

The term "resolution" refers to the output quality of a photo or graphic image (in other words, what it looks like on a computer screen or on the printed page). Different measures are used to describe the resolution of an image, depending on what medium is being used to view it. On a printed page, resolution is measured in "dots per inch" (d.p.i.), since images on almost all printed materials (books, magazines, leaflets, and scientific journals) are made up of thousands of tiny dots of ink. You can use a magnifying glass to examine a photo in a magazine or a figure in a journal (figure 8.1a); the dots that make up newspaper images are quite large and can often be seen with the naked eye. Good quality printing usually requires images that are 300 d.p.i. For an image on a computer screen, resolution is described in terms of "pixels per inch" (p.p.i.), since a screen image is made up of thousands of pixels. On screen, 72 p.p.i. will look fine. However, journals may require online figures to be at a higher resolution for optimum quality.

The important thing to bear in mind for print journals is the size of the final image as it will appear on the journal page. A photo that is made up of a single one-inch square containing 300 × 300 dots will look great if printed on the page *at that size*. However, if the picture is to be displayed

a

b

FIGURE 8.1 (a) Photos and graphics printed in books, journals, and magazines are made up of a "screen" of ink dots. A view at a high magnification (inset) shows the dot pattern that creates the picture. (b) Pictures on a computer screen are made up of pixels. At a high resolution (left side) these are invisible, but if a low-resolution image (right) is shown at a large size, the pixels become visible.

on the page at a width of four inches, it will looked unfocused, since the same 300 × 300 dots are now spread out more widely. Similarly, if you have an onscreen image that is one inch in size, made up of 72 × 72 pixels and you increase its size on screen, you are just making the existing pixels bigger so that you start to see them as jagged lines making up the image (figure 8.1b).

The instructions to authors of print journals will usually specify whether they want low-resolution files (small file sizes, usually of 1 MB or less) at first submission or whether authors are expected to provide high-resolution versions (large file sizes, of 2 to 5 MB or more) straight away. Small files are quicker to open and take up less room on a computer hard drive. Peer reviewers often prefer to work with low-resolution files for the sake of speed and convenience. However, when the journal is printed, a high-resolution image will be required. If the journal is an online-only publication, then medium- and low-resolution photos and graphics will look fine, although the higher the resolution the better the quality, even on the computer screen.

If you have a low-resolution image, there are some software packages that may help you to enlarge the size with only minimal loss of quality, but only up to a certain point. The best course of action is to ensure that you have high-resolution images to begin with. Adjust your camera to a high-resolution setting or, if using a graphics package, make sure you reproduce or render the illustrations at a high resolution. One other alternative you can try, for graphics and photographs, but *not* for photos out of journals or books (figure 8.2), is to create a good quality print of the figure and then rescan it at a higher resolution.

Check the instructions to authors to find out the file types and resolution required for figures. If you are submitting to a print journal and you are providing low-resolution figures for peer review, make absolutely certain that you have, or can easily obtain, the high-resolution versions. Researchers taking photographs during fieldwork sometimes make the mistake of adjusting their digital cameras to take pictures at a low resolution so that more pictures will fit onto the camera's memory card. Unfortunately, low-resolution picture files cannot be used in print journals, even though they may look fine on a computer screen; if the original photo is low resolution then not much can be done to improve this. Do not try to save these low-resolution photos at a higher resolution—this simply creates high-resolution, low-quality images that will look awful in print. If you are taking photos that may one day be published, whether you are out in the field or in your lab, make sure the camera is adjusted to take high-resolution photos.

FIGURE 8.2 A moiré pattern, caused by the "screen" pattern of dots from the first printing interfering with the pattern of dots from the second printing. This will only appear in the reprinted photo.

Do not use scanned material

Never submit photos or graphics that have been scanned from a book, journal, or other printed material. First, this is illegal without proper permission from the original publisher (see sidebar 8.2 on transfering copyright and sidebar 8.3 on the permissions process). In addition, if you submit materials that have been scanned from previously printed material and submit it to another journal or print publication, there is a serious danger that when it is reproduced on the printed page it will appear with an ugly interference pattern all over it, known as a moiré pattern (figure 8.2). This effect can only be seen in the final printed version, so the Editor will not know what you have done until it is too late. This is one way to make yourself very unpopular with that particular Editor.

Check copyright and permissions

Copyright law is extremely complex; indeed, whole books have been written on this subject. What we cover here are simple but absolutely vital rules that you *must* follow. Merriam-Webster's Dictionary defines "copyright" as "the exclusive legal right to reproduce, publish, sell, or distribute the matter and form of something (as a literary, musical, or artistic work)." As the author of an as yet unpublished manuscript, you hold the copyright to your own work; in other words, you have the right to publish it, sell it, distribute it, or use it in any other way you want. However, when your manuscript is accepted by a journal, you are usually asked to sign a document in which you consent to pass ownership of copyright to the journal's publisher. Some of the newer types of publications, such as open access journals, are an exception to this rule (see sidebar 5.1 on open access and the section below on Creative Commons licenses). Unless specified otherwise, the copyright to the entire content of the journal rests with the publisher—this is normally nonnegotiable. (See sidebar 8.2 on transfering copyright).

SIDEBAR 8.2

Why transfer copyright?

ERIC S. SLATER, ESQ.
Senior Manager, Copyright, Permissions & Licensing, American Chemical Society

Most scientific and scholarly publishers require that authors transfer copyright to them as part of the publication agreement between the publisher and authors. In this context, the publisher becomes the rightsholder for purposes of copyright and, in return, many publishers grant a number of rights back to authors (more on this below). It is important that authors understand why they are required to transfer copyright, which includes the legal formality involved in the process. As discussed in sidebar 5.1 (Why Choose Open Access), some publishers do not require the transfer of copyright, but at present most scientific publications do have this requirement.

First, with respect to the legal aspects, under US Copyright Law, it is a requirement that copyright transfers be done in writing. To facilitate this process, most publishers have forms for transferring copyright (sometimes referred to simply as a "copyright transfer agreement," a "publishing agreement," or something similar). Most publishers require authors to submit a completed form (meaning a fully executed or signed form) before they will begin production or schedule publication of a manuscript. It is therefore crucial that authors fully understand what is required by the publisher at the outset.

Second, authors are not always clear as to why a publisher requires the transfer of copyright. Although transfer of copyright to the publisher is the standard practice in scientific and scholarly publishing, this still does not really explain why *it is necessary. The following list highlights some major reasons for transferring copyright, reasons that hold true for both for-profit and not-for-profit publishers and societies:*

- ***The prestige factor.** While the Internet has made it relatively easy for authors to "self-publish," it is more advantageous from a professional standpoint to publish with an internationally known and well-respected publisher in the author's particular industry or field of study. In scientific publishing this would entail publishing with organizations such as the American Chemical Society or the Royal Society of Chemistry, or commercial publishers such as Elsevier or Wiley-Blackwell, to name a few.*
- ***The permissions process.** The publisher provides a single, central contact for those who are requesting permissions or licenses. Those activities requiring such permissions include, but are not limited to, photocopying full articles or excerpts of articles, the reuse of previously published figures and tables, and the use of abstracts or other contents of a published paper. This process allows for a consistent policy to be enforced and ensures efficiency in assisting a publisher in fulfilling its mission to make the work available to the widest possible audience. At the same time, the burden of handling such requests is removed from the author, and the publisher can ensure that only legitimate requests and uses are approved. The permissions process has become much easier in the Internet age, mainly due to the Copyright Clearance Center's (CCC) RightsLink service; this essentially allows for a one-stop process and virtually instantaneous permission as many publishers have partnered with CCC to permit use of content.*
- ***Protection against copyright infringement and other legal protection.** Publishers are best positioned to protect the author's work, both domestically in the US and internationally, against copyright infringement, libel, or plagiarism. The publisher has the resources to protect the work against unauthorized copying and distribution, and has the ability to enforce these rights so as to prevent unauthorized use. This would likely involve the sending of cease-and-desist notices and takedown letters to stop unauthorized use once discovered and to enforce publisher-centric ethical guidelines, among other things.*
- ***The final version of the paper.** The publisher is in the best position to ensure that the final version of a manuscript is the "official version of record." Publishing on the Internet potentially allows any number of versions or drafts of a manuscript to be posted; in this context, the publisher becomes the official source of what represents the final version of a paper.*

- ***The integrity of the work.*** *The publisher, whether this is a scientific society or a commercial publisher, accepts the responsibility for promoting the integrity of the published work, as described above. The publisher takes on all of the tasks associated with the publication of the work from start to finish, including the peer review process, editing, production and layout, and dissemination of the work. The publisher should also be positioned to market and promote it.*

Finally, it is important to note that most major publishers will grant numerous rights back to authors, allowing them to use and exploit their written works on their own. Publishers are concerned with and fully recognize the rights of authors from this perspective, and will typically permit the author to reuse the work in a wide variety of ways (including but not limited to reprinting the work, creating derivative works, distributing the work to colleagues, and more). This also extends to the employers of authors, when works are written under a work-for-hire scenario. Editors and authors alike should make certain that they fully understand any such agreement by reading it carefully before signing, and should consult with legal counsel if necessary.

Copyright on figures and tables

If you take photographs or create graphics to illustrate an article, the copyright on the images also passes to the publisher when the paper is published. However, if, for example, you work for an organization that sells its photos online, you may be able to negotiate with the publisher to retain the copyright on those photos. If the publisher agrees, the copyright sign, followed by your name (as the photographer) or the name of your organization, will appear in close association with the figure itself, when the journal publishes it.

If you wish to use a photo or graphic that has already appeared in another publication, it is your responsibility both to find out who owns the copyright for that item and to get their written permission to use it as part of your paper (see sidebar 8.3). Copyright and permission policies vary among journals and publishers, so you need to check carefully each time you come across this issue. Don't assume that what was true for a previous publication will also be true for a new one. You will need to pass that written permission to the journal in which you are publishing, so they know they can legally republish that item. To find out who owns the copyright on a photo or graphic, carefully examine the figure and its accompanying caption in the original publication. Look for the word "copyright" or a copyright sign (©) somewhere in the figure caption or along one edge of the figure (examine the top, bottom, and both sides). If you cannot find this next to the figure or in the caption, then it is probable that one of the authors took the photo or created the graphic and the copyright passed to the publisher. If this is the case, you will find the copyright sign, linked to the name of the publisher of the journal or to the name of the journal itself, somewhere on the page, or on one of the early pages of the journal. In books, copyright information may appear on a page near the very beginning or somewhere near the back.

SIDEBAR 8.3

The permissions process

CAROL EDWARDS
Publishing Manager, TESOL International Association

As you read an article in a journal, say, TESOL Quarterly, *you may see wording like this: "Table adapted from Table 6, in Davison (2004), copyright © 2004 by SAGE Publications. Reprinted by permission of SAGE." This statement tells you that the table is copyrighted by another publisher (SAGE) and that the* TESOL Quarterly *author has received permission from SAGE to reuse it.*

When you write a paper for a journal, or any publication, you must give careful consideration to any content that is not your own original creation, including text, tables, figures, charts, photos, and drawings that you want to quote, adapt, or use in the format in which they originally appeared. If the material is copyrighted, you must go through the process of getting permission to use it in your paper. The process is not difficult, is legally required, and is necessary to allow your publisher to get your paper into production without delay. Note, however, that you do not need to request permission to use or adapt material from an open access article, as long as you give the original author appropriate acknowledgment for the work. (See sidebar 5.1 for more about the open access option.)

WHO OWNS COPYRIGHT?

To determine who owns copyright, look for a copyright sign (e.g., © 2004 SAGE) in the publication where the table, figure, or graphic originally appeared. If you are considering using such materials from a publication, look carefully for the copyright symbol, particularly at the beginning of the book or issue. An individual artist or photographer may mark their material, for example, "Not to be copied without permission" or "Do not use without consent."

GETTING PERMISSION FOR REUSE

The first step towards getting permission is to contact the copyright holder; in the example above, the copyright holder was SAGE. Every copyright holder should list a Permissions Manager or similar person to contact, either within the publication or on a website. Each copyright holder has its own rules for granting permission, so you cannot assume that the process you follow for one copyright holder will work for all the others. The copyright holder may ask you where the material will be published, how many copies will be circulated, and other questions. It is important to contact the copyright holder as soon as you know you want to use the material in your article because the permissions process can take months and ideally should be finished before you submit your own paper for publication. Otherwise, your article may be delayed and may have to go into a later issue of the journal.

DOES PERMISSION COST?

When you contact the copyright holder, you may find there is a fee for using the copyrighted material. Most publishers put the burden on the author to pay any fees required. Alternatively, some copyright holders will have a policy that will allow you to use a portion of the material at no cost. For example, you may be allowed to use up to 150 words with no permission or fee involved. However, this policy varies for each copyright holder, so you must check each time you want to use previously published material.

ASK THE CORRECT RIGHTS HOLDER

When the original creator of the material has transferred copyright to the publisher via a signed document, the publisher, not the original author of the material, owns the copyright. You need to be sure you ask the legal owner of the copyright for permission to reuse the material. To learn more about transferring copyright to the publisher, see sidebar 8.2 on Why Transfer Copyright?

WHAT HAPPENS IF I CANNOT GET PERMISSION?

If you do not receive permission to use copyrighted material, the copyright holder can force the publisher to take the material out of circulation. This means an online article will be taken off the Web and a print publication will no longer be available to subscribers or libraries.

MORE INFORMATION

In the US, the Copyright Clearance Center (www.copyright.com) manages the permissions process for publishers and users. The pamphlet Copyright Basics *(http://www.copyright.gov/circs/circ1.pdf) clearly explains copyright and the law. Copyrighted material is protected by the Copyright Act in the US and by similar laws in most other countries as well.*

You must obtain written permission from the named individual or organization that appears immediately following the word or sign (e.g., Copyright 2010 Smithsonian Institution or © M. Baker). It is *not* sufficient to get permission from the author of the paper or book chapter in which the figure appeared, unless it is his or her name that appears next to the copyright sign. However, it is highly advisable to obtain permission from the author as well, not for legal reasons but as a courtesy. You could damage your reputation among researchers in your field if word gets around that you are using other people's figures without asking.

You also need to find out whether there is a fee for republishing the photo or graphic. Many publishers will charge you for reproducing a figure, table, or any other part of one of their publications, as will some nongovernmental organizations. If it is discovered *after* publication that you have published a figure or other copyrighted material without permission and without the required payment, this may be a serious matter, with possible legal repercussions.

If the figure you want to use is in a journal, consult the journal website to obtain the contact details of the Permissions Editor or of the permissions department of the publisher. Contact them as soon as possible, giving all the details about the figure (name of publication, volume, issue, and page number, name of author, title of paper, and figure number) and also explain where and how you want to reuse it.

Once you have received the necessary permission, make sure that you clearly acknowledge the source of the material and the copyright details in the figure caption. The original publisher will often provide instructions on how they want to be acknowledged in your paper. If not, there are some standard styles you can use, such as: "Reproduced by permission of [name of journal]" or just "© Thomson Reuters 2010."

Because permissions departments can be very slow to respond to requests, don't wait until your paper has been accepted for publication before you contact them. Once you have received written permission to reproduce the photo or graphic, send this to the Editor, together with your manuscript, as part of the original submission, or as soon as possible thereafter.

The rules for reproducing a table or any other material that has already appeared in another publication are the same as for figures. Obtain written permission to reuse the table and submit this to the journal you wish to publish in, just as you would for a photo or graphic.

Photos and photolibraries

Many print journals charge a fee for the use of color in figures. However, this is not true of many online-only journals, so if you are publishing in an electronic publication, you may not have to pay to include color images. In many cases, and particularly when the manuscript deals with very specialized research, you will have to provide these images yourself or ask a colleague or another author for something suitable. However, if you need high-quality, high-resolution images of landscapes, animals, plants, or people, there are a surprising number of websites where these can be found. Appendix 4 provides a list of websites where you can find a wide selection of low-cost or free images. You probably won't be able to get highly specialized or technical photos, but it is well worth investigating some of these sources, as there is an amazing variety of material available, if you know where to look. It is useful to know where to obtain good photographs, not just for papers, but also for use in PowerPoint presentations, as well as reports, websites, or blogs.

Many US federal agencies have their own photo libraries, and much of the material in these is in the public domain (not protected by copyright) and can therefore be used free of charge. Images with filenames that end in ".gov" often fall into the category of public domain material. These agencies require only that you acknowledge them as the original source of the photo. Some make high-resolution versions directly available for downloading, while others only post low-resolution photos online, so you will need to contact the relevant agency to get the high-resolution version. However, if you find an image you want to use on one of these websites, always check the website carefully and make sure there are no restrictions on what you intend to download and use. If you find a notice on the website that says "Copyright © 2011 US Geological Survey. All rights reserved," then obviously you need permission to use the images.

Internet material

The Internet is full of beautiful photos and graphics; in fact, there is so much material out there that you are bound to find something on somebody's website that will fit perfectly in your paper, and so you may be tempted to just download it straight from the Internet and use it. You must *not* do this for two reasons:

(1) Unless stated otherwise, it is illegal. Much of the material on the Internet is protected by copyright in just the same way as it is in a journal

or book, so you cannot use it without permission. Of course, you can contact the organization or individual who owns the material and get permission, in the same way as described above.

(2) Photos and graphics on the Internet are usually at low resolution (72 d.p.i.), unless stated otherwise and are therefore not suitable for use in a print publication, which requires much higher resolution (normally 300 d.p.i.). Just because a photo looks wonderful on a computer screen does not mean it is usable in a print publication.

Creative Commons licenses

As discussed earlier, whenever you create something, whether it is a scientific paper, a graphic, or a piece of artwork or music, you automatically own the copyright to that piece of work. All rights to the work belong to you. If anyone else wants to use it, they have a legal requirement to get your permission first. However, if you do want people to reuse your work, a relatively new form of copyright licensing is now available for this purpose, called the Creative Commons (CC) license (http://creativecommons.org).

Although a number of well-known entities and organizations use CC licenses (e.g., Flickr, Wikipedia), the majority of scientific journals do not. The exceptions to this are some of the open access journals, such as those published by the Public Library of Science (PLoS). Unless your chosen journal already uses CC licenses, do not expect to be allowed to do so. If you want your paper to be free for everyone to see and use, you will need to publish it in one of the new generation of open access journals (see sidebar 5.1 on open access).

SIMULTANEOUS SUBMISSIONS

Some inexperienced authors become discouraged after they have been rejected by multiple journals and so decide to save time by sending the same paper to two or more journals at the same time. *Do not do this.* If the Editors of these journals find out—and they very likely will—your paper will most probably be rejected immediately. The Editors involved may even refuse to consider manuscripts you send to their journals in the future.

Submitting manuscripts simultaneously to multiple journals wastes the Editor's time and that of the peer reviewers. Furthermore, since the number of peer reviewers within a particular field of study may be relatively small, if you send your paper to two journals, it is quite possible that both Editors will send it to the same reviewer. That reviewer will quickly contact the two Editors to tell them what has happened. Alternatively, if your paper is accepted by two publications, you will need to withdraw the manuscript from one of them. In either case, everyone will know that you have broken one of the basic ethical principles of scientific publishing.

THE BOTTOM LINE

If your science is unoriginal or poorly executed, this book probably cannot help you much. However, if the science is sound, and will make a clear contribution to the collective knowledge in your field, then don't give up. Even if your paper has been rejected, tailor your manuscript to fit each new journal's requirements, do all your preflight checks, and send the paper out again. Remember that getting your paper into peer review counts as a success, even if it is later rejected. You will have the peer reviewers' comments to help you improve the paper or the science it describes. Eventually, it will get published.

CHAPTER NINE

Who does what in peer review

Peer review is such a fundamental element of critical scientific thinking that the entire scientific and scholarly community should arguably take on the responsibility for improving and maintaining its quality—a major, long term commitment.

FRANK DAVIDOFF, Editor Emeritus, *Annals of Internal Medicine BMJ.* 2004. **328**: 657 doi: 10.1136/bmj.328.7441.657

LOOKING INSIDE THE BLACK BOX

For some authors, sending their manuscript off to a journal and then receiving a decision letter some time later is rather like pushing their paper into a black box and waiting for something to come out at the other end. They have no clear idea of what happens inside the box. Who is making the decisions? How are those people chosen? Can anything be done to hurry the process along? The aim of this chapter is to lift the lid off the black box and take a look at what goes on inside.

Peer review—which in the context of this book is the consideration of your manuscript by a number of experts in the same field—is an integral part of the process of science publishing. Having your work criticized and, in some cases, rejected as substandard can be a painful ordeal that every scientist must face, but the end result is that when you look at a paper in a peer-reviewed journal you know that one, two, three, or even four subject matter experts have considered it, requested any necessary changes, and now deem it to be scientifically sound.

In the past, authors typed out their manuscripts, made two or three copies (as specified in the journal's instructions to authors), packaged them all up, and posted them to the journal offices. Nowadays, you are more than likely to find yourself grappling with an online submission system, which acts as a web interface for everyone involved in the peer review process. These systems allow authors to submit manuscripts electronically and track them as they move through peer review, while also allowing Editors, editorial staff, and reviewers to upload and download manuscripts and reports, and view the status of articles (see sidebar 9.1 on Online Manuscript Submission and Peer Review Systems).

SIDEBAR 9.1

Online manuscript submission and peer review systems

LYNDON HOLMES
President, Aries Systems

Peer review systems are like plumbing for scholarly publishing. They manage the flow of scholarly manuscripts from submission to acceptance and beyond. Like good plumbing these workflow systems should be invisible and reliable, and they should afford some luxury.

While most of us could theoretically do our own plumbing, we usually discover that it's better to have a professional build and maintain the system. It's the same with online peer review systems. Most scholarly societies, publishers, and university presses could develop and host their own workflow systems but have discovered that it's easier and more efficient to adopt a commercial solution.

The evolution from paper-based workflows to electronic processes began in the 1970s, with DOS-based systems, and those paved the way for the Windows desktop systems and web-based peer review that were popular in the mid-1990s (Tananbaum and Holmes 2008). Today's peer review systems are the result of tens of millions of dollars of investment, and leading vendors typically invest in excess of $5 million a year to keep their web-based platforms at the peak of functionality and up-to-date.

Today, the vast majority of scholarly journals offer online submission and peer review, using one of a handful of commercially available systems including Bench>Press, Editorial Manager, eJournalPress, and ScholarOne. Some organizations use open-source software, but these systems cannot compare with those that are developed and maintained by full-time professionals.

SERVING THE AUTHOR

One of publishers' key objectives in using an online peer review system is to improve their services to authors. For example, most systems allow authors to quickly check the status of their submission, review prior correspondence, or review online submission instructions that will help them meet editorial and mechanical submission requirements.

Some publishers configure their system to provide early, automated feedback to authors concerning the quality of their manuscript. For example, Springer SBM's system automatically links the author's submitted bibliography to PubMed and CrossRef. This is a valuable service to authors since it provides early warning of bibliography problems and suggested responses. Similarly, some peer review systems automatically compare the characteristics of uploaded image files to the journal's publication standards. Again, the author gets an early warning of any image corrections they need to make to move their submission forward, towards publication.

Publishers can also use their peer review systems to help them quickly find reviewers for papers. They do this by maintaining subject domain keywords in a structured hierarchy and encouraging authors and potential reviewers to identify their area of expertise from the keywords in this list. The system can then match authors and manuscripts to potential reviewers.

Authors can expect to see further innovations in the interface and functionality of manuscript submission systems that will make the submission process easier and more responsive to their needs. For example, leading systems now allow a language "toggle" so that authors can view the interface in their own language, including simplified Chinese, Japanese, French, and German.

WHAT AUTHORS NEED TO KNOW ABOUT USING PEER REVIEW SYSTEMS

While most peer review systems are developed and maintained by vendors, many specific aspects of the workflow are highly tailored and customized to meet the specific requests and requirements of each individual journal. For example, while most peer review systems can technically accept submissions in PDF format, some do not allow this because once a paper is accepted, Copyeditors want immediate access to the original files for editing with the goal of speeding time to publication. This illustrates how journals try to balance the need for up-front author convenience with meeting an important downstream objective of early publication.

To take full advantage of the power of manuscript submission systems and keep the Editors who receive the manuscripts happy, authors should:

- *Carefully read and comply with the online instructions included at each stage within the manuscript submission system. Journals and systems designers have worked hard to tailor the systems to collect information from authors that Editors and production staff need to keep their production workflow moving forward with the greatest efficiency. In other words, all the steps the authors must follow and all the information the system asks for are necessary, and authors should not skip or overlook anything.*
- *Correspond with the editorial office through the peer review system rather than through email once your manuscript has been submitted. The system*

creates records of all communication, and both authors and editorial staff can easily access these messages. Using the tools within the system helps ensure that no important information is lost.

- *Explore and make full use of the many useful features that well-designed manuscript submission systems provide—these allow authors to check the status of submitted manuscripts; inform the editorial office of times when they'll be unavailable; register temporary email addresses; integrate with bibliographic searches; use automated formatting and quality checks.*

Some systems also provide features that are useful after papers have been accepted, such as allowing authors to update files or review and submit final page proof edits.

THE FUTURE

Changing needs and technologies (such as new browser capabilities) will allow authors to benefit from many innovations that will be introduced during the next few years. New tools such as scholar identification systems (e.g., Open Researcher & Contributor ID, ORCID; www.orcid.org) may help address the chronic author need for a single sign-on between different journal and publisher systems. The widespread adoption of tablet computers, especially the iPad, by scholars may facilitate new ways to interact with peer review systems that go beyond the browser interfaces currently supported on tablet computers. From a technical point of view, authors can look forward to greater speed, convenience, and access options during scholarly manuscript submission and peer review.

WHO'S WHO IN THE PEER REVIEW PROCESS

The first place to look for information on who is deciding the fate of your manuscript is on a page near the front of the journal, known as the masthead page. In larger publications, and those with numerous issues per year, masthead information will take up a whole page or more. In smaller journals, it may just take the form of a small panel. The masthead usually lists the permanent staff in the journal office, including editorial, production, marketing, and other staff.

When it comes to peer review systems, every journal is different. In some cases, permanent editorial staff play key roles in the process, in others all the administration, reviewing, and decision making is done by "volunteers"—usually scientists who do this work in addition to their regular day jobs as researchers, lecturers, or other science-related employees. The titles given to the people who fill different positions in the journal hierarchy also differ widely among journals, and this variety can confuse novice authors or others who are trying to figure out who they should talk to in order to get an answer to a particular question. Big weekly or monthly publications usually have a full-time Editor-in-Chief or Editor and a large permanent staff; these larger publications also use a small army of academic volunteers, usually known collectively as the editorial board. A small annual, semiannual, or quarterly journal, on the other hand, may have just one Editor, who must try to identify relevant reviewers for each submitted manuscript and make most if not all the decisions, with or without secretarial or editorial support. Editorial structures may also differ, depending on whether the journal is published by a not-for-profit academic society or association or by a for-profit commercial publisher. For more detailed descriptions of the different individuals who work in, or closely with, the editorial office, see the section on journal staff and volunteers later in this chapter.

Finally, journals differ in the degree of anonymity that characterizes their peer review process (see the section below on different forms of peer review).

Peer reviewers

Peer reviewers are subject matter experts who are invited by the journal Editor or staff to review papers on subjects about which they are considered to be very knowledgeable. They can refuse, of course, if they are too busy or if they feel that they are not sufficiently knowledgeable about a particular aspect of the topic. Unfortunately for editorial staff, refusals are fairly common, so that it can take several days, or even weeks, to find the necessary one, two, or three (depending on the journal) suitable peer reviewers for each paper.

Once a peer reviewer has accepted the task, he or she is expected to read the paper carefully and write a detailed report, not only pointing out any flaws and omissions in the work but also making thoughtful, constructive suggestions on how to improve the manuscript. It is important to note here that peer reviewers don't make decisions about the papers they review; they can make recommendations, but the final decision regarding the fate of the manuscript does not rest with them, but with an Editor or Associate Editor.

A good peer reviewer will often be in the database of a number of journals and will be called on frequently to review papers. They usually have to do the reviews in their spare time, which means evenings and weekends, and they normally get no reward and little thanks. They do this work because it is an expected part of a scientist's career and because they hope that someone else will do the same for their papers. Of course, a few individuals "cheat" the system; they always find a good reason for not accepting the papers they are offered: "too busy," "heavy teaching load this semester," "not really my area of expertise," "just leaving for an extended fieldtrip." Worse still, from an Editor's point of view, is the reviewer whose report is five weeks overdue, and who finally sends back a three-line report, stating that the paper is good, bad, or indifferent but providing no detailed guidance for the author regarding how to improve the manuscript.

As an author, however, you need to remember that even the most conscientious reviewers may be tired, stressed, or in a bad mood when they finally find the time to sit down with your paper. They may get irritated by poor presentation, poor English or, worst of all, poor science. As a result, reviewers are sometimes not as tactful or as polite as they should be when writing their reports (see chapter 10, "Dealing with Decision Letters").

When you have the opportunity to suggest peer reviewers for your paper, you should mention people whose area of expertise qualifies them for this role and whose work you respect. However, there is no point in suggesting very high-profile scientists, since they probably won't have time to do the review. You should also avoid recommending former supervisors or close colleagues and friends, or anyone that could be perceived as having a conflict of interest. In other words, don't recommend individuals whose past or current professional or personal relationship with you might bias their judgment of the paper.

PEER REVIEW: A FLAWED SYSTEM

A well-known and much-repeated saying in the scientific publishing world is that the peer review system is flawed, but it's the best we've got. This is perfectly true. Peer review is carried out by human beings who cannot help but bring their own beliefs, biases, and prejudices to the task. Even so, the vast majority of Editors and peer reviewers take their task very seriously and do their best to make a balanced and fair assessment of each paper and to provide the authors with useful feedback.

Over the years, various author groups have felt that they were at a disadvantage when it came to publishing their work, and were convinced that Editors were biased against their submissions, even before reading the abstract. This belief has led to many published papers on "attitudinal biases" in science publishing—that is, biases that influence editorial decisions or citation rates and that have nothing to do with the scientific merits of the papers. Factors that have been investigated include gender bias (are papers with male first authors more likely to be accepted for publication or more frequently cited than papers with female first authors?), language bias (are papers whose first author is a native English speaker more likely to be accepted for publication or more frequently cited than papers whose first author is a nonnative English speaker?), and bias against country of origin (are papers from certain countries less likely to be accepted for publication than those from other countries?).

Many women have felt at a disadvantage in terms of gender bias in the peer review system and non-English-speaking authors have also felt that they are at a major disadvantage when trying to publish papers in English-language journals. The accusation of language bias in particular cannot entirely be denied. Editors do not like to admit it outside their own circle of colleagues, but their hearts sink every time they note the arrival of a new manuscript, written by a nonnative speaker of English, and see almost incomprehensible text that will be a struggle to read and which they suspect (not necessarily correctly) will yield badly designed experiments, dubious results, and unfounded conclusions.

If you are a nonnative English speaker and you want to get your work into the international arena, getting your papers past the English-language-based peer review system remains a challenge that must be overcome (see sidebar 4.2, "The Challenges of Publishing as an International Author," and appendix 1 for a list of resources that may prove helpful).

Editors and publishers are extremely aware of the dangers of bias and other inequities within the peer review process, recognizing that a fair and effective system is a fundamental part of the development and dissemination of new scientific information. To address the potential inequities in the review system, about twenty years ago Editors of the *Journal of the American Medical Association* and the BMJ (British Medical Journal) Group formed a collaboration to launch an international congress on peer review. This meeting has since taken place every four years. The Congress focuses exclusively on the results of research on peer review and on ways to improve the peer review process. Many hundreds of publishers and researchers attend these meetings and take the information reported there back to their workplace to try to improve their publishing practices. Extended abstracts or full text of most of the papers presented are freely available on the Congress's website at www.ama-assn.org/public/peer/peerhome.htm.

DIFFERENT FORMS OF PEER REVIEW

There has been much discussion and debate, among Editors and within the scientific community in general, regarding the different models of peer review and which one is least prone to biases and other inappropriate influences. Most journals use one of three models of peer review: single-blind review, double-blind review, or open or community review. Each of these models has advantages and disadvantages, so that the system used by each journal depends on the beliefs and biases of the publisher, the Editor-in-Chief, or the Editorial Board.

Single-blind review

The single-blind model is the most commonly used among life science and biomedical science journals. In this system, the peer reviewers remain anonymous to the authors, but reviewers are able to see the names of the authors. Reviewers can therefore be openly critical about a scientist's work without any fear that the recipient of these criticisms may, sooner or later, retaliate in some way. Many reviewers prefer a single-blind system since it can be embarrassing to meet colleagues or future collaborators after you have heavily criticized their latest manuscript. Proponents of the single-blind system believe that if the reviewers' names were visible to authors, those reviewers might end up providing weak or incomplete reviews, because they would be afraid of causing offense or making enemies.

When Associate Editors (see "Journal staff and volunteers" later in this chapter) play a role in decision making, they are not normally anonymous and often correspond openly with authors on behalf of the journal. Associate Editors have less of a problem with being out in the open since a lot of the criticisms of the manuscript will be coming from the reviewers, leaving the Associate Editor to put these into a constructive framework, sometimes adding further comments and suggestions, as well as presenting possible options for the authors to consider.

Double-blind review

In the double-blind review system, the reviewers' names are hidden from the authors and the authors' names are hidden from the reviewers. Proponents of the double-blind model point out that a particular author's name on a paper (e.g., a very senior researcher) or a geographical location (e.g., a country whose authors have a reputation for plagiarism) may influence the opinions of the peer reviewers, either positively or negatively. The presence of a highly respected author at the top of the paper may lead the reviewers to be more lenient or, conversely, to judge it by a higher standard than if it were a first attempt at publication by a young postgraduate student. A name associated with a particular nationality or problem may also engender positive or negative biases (e.g., reviewers from the same country as the author may be tempted to help out their countryman, or an Associate Editor may react negatively to a paper from an author with whom they had difficulties in the past), leading to inequities in the peer review process.

One of the main arguments against the double-blind system is that since science is quite a "small world," unless you can completely eliminate all possible clues to the authors' identity within the text (which is often hard to do), the peer reviewers may be able to make a good guess at the identity of the authors. A specific approach to the research, a particular study area, or a certain technique can all act as clues. If you are working in a fairly specialized field, you are probably familiar with most of the other researchers in that field and may know many of them personally, as well as being familiar with their work. Peer reviewers may waste time studying these clues in an attempt to guess the identity of the authors. Having arrived at a satisfactory conclusion (whether right or wrong), they may then proceed to subconsciously apply all the biases they would have done had everyone's name been visible in the first place.

Open peer review or community review

As the name suggests, in open peer review neither the authors nor the reviewers are anonymous; instead, everyone's name is visible. This transparency often means that reviewers will be more tactful and polite, since they have to be ready to justify their criticisms the next time they meet the author whose paper they recently slammed. Unfortunately, they may also choose not to accept the task of reviewing the paper at all, which makes life more difficult for the Editor.

Community-based peer review is an extension of open peer review, in which papers are posted online and the scientific community is invited to comment on and criticize the paper. In these systems, both authors and those posting comments are known to each other. The authors can respond to the comments and, in theory, will be able to improve the paper to a high enough standard for publication through this open dialogue with peers.

Probably the best example of community peer review can be seen in some physics, mathematics, and computer science journals, where this system has been in use for many years. As soon as a manuscript is ready, the author may choose to submit it for posting as a preprint on Cornell University's arXiv (pronounced "archive"). Editors check to make sure that it is of a suitable scientific standard, but otherwise make no judgments about the manuscript. Members of the scientific community are free to provide feedback on the paper, which is then sometimes (but not always) submitted for publication to a peer-reviewed journal. Otherwise, the paper remains on the arXiv site as an "e-print."

Different disciplines and various publishers, including the Public Library of Science, have instituted variations on the theme of open and community peer review, sometimes termed "interactive public discussion," but these systems are still rarely found among the life sciences and medical journals. In 2006, *Nature* invited authors of newly submitted papers to take part in an experiment (Anonymous 2006), in which they posted papers on a public server and invited signed comments from the readers, while at the same time carrying out normal, single-blind peer review on those papers. During the four-month trial, the authors of 71 papers (out of a total of 1,369) agreed to take part. Of these, 33 (46%) received no comments at all, while 38 (54%) attracted a total of 92 comments. Of these, 49 comments were directed at 8 papers. Although there was quite a lot of interest in this experiment at the time, *Nature*'s Editors were disappointed by the small number of authors who agreed to participate and the small number of comments received on the papers that were entered into the trial. Perhaps not surprisingly, the Editors decided not to adopt open peer review.

DECISION MAKING AND SPLIT DECISIONS

Authors sometimes wonder what happens when there are two peer reviewers looking at a paper and they disagree with each other. In other words, what happens when one reviewer suggests that the paper requires only minor revision to be acceptable for publication while the other feels that the paper is deeply flawed and should be rejected or, at the very least, that it requires an extensive overhaul?

In such cases, Editors basically have two options. First, if they are experts in the subject, they may read the paper again and cast the deciding vote themselves, adding extra comments of their own if they feel that the two original reviewers have missed important points. If they are not sufficiently familiar with the topic, they may solicit a third review and let this decide the issue.

Unfortunately, no peer review system is perfect and, being human, Editors and peer reviewers do make mistakes, although not as often as aggrieved authors, staring at their rejection letters in anger and disbelief, might think. You need to study the reports on your papers carefully and then try to look objectively at your work. Invite colleagues whose opinion you trust to do the same. If you still feel that there has been a serious error, see chapter 10 on how to appeal the decision.

PEER REVIEW: WHAT'S IN IT FOR ME?

If peer reviewing is so time consuming, then why, you might ask, should you bother? What do you get out of it? The answer is "quite a lot." First and foremost, as mentioned above, peer reviewing is a recognized part of being an engaged and successful scientist. It is also a matter of quid pro quo—you will be doing this service for other scientists in the hope that they will do the same for you. Having a line on your résumé saying that you serve as a regular reviewer for one or more journals in your field can be helpful when you are seeking a new job or a promotion. Serving as a peer reviewer also provides you with the opportunity to see the latest results from your fellow researchers. That being said, every paper you see as a reviewer is confidential and privileged information, and you cannot use the knowledge you gain in any way until after it has been officially published.

Peer reviewing is also an opportunity to exercise your critical faculties and to think about what makes a good paper (or a bad paper). Exercising these skills can help to improve your own writing and may also make you more familiar with the requirements of the journals you are reviewing for. This knowledge may, in turn, help you when you are considering these journals for your own papers.

Finally, good peer reviewers are worth their weight in gold to Editors. If you do a good job reviewing for a journal in your field, you may eventually find yourself being invited onto the editorial board. This will put you on good terms with Editors and staff of the journal, which can also give you a small advantage when submitting your own papers. When Editors see a familiar name at the top of a newly submitted manuscript (a name they associate with excellent critical thinking skills, thoroughness, and attention to detail, and punctuality in returning work), well, let's just say that it can't do you any harm if the name they recognize is yours.

JOURNAL STAFF AND VOLUNTEERS

As mentioned earlier, authors do sometimes wonder who exactly is making decisions in the editorial office (or elsewhere) and how they got to be in a position to do so. Understanding a little more about the main players in the editorial process can help answer these questions.

Editor-in-Chief or Editor

The buck has to stop somewhere and it is on the Editor-in-Chief's (EiC) or Editor's desk that it usually comes to rest. Although many of the biggest and best-known journals in the world have full-time, professional EiCs or Editors, for the great majority of publications that you will be submitting papers to, the EiC is a practicing academic, involved in research, teaching, or high-level administration, and is often a big name in the scientific field that the journal covers. This individual is usually responsible for the overall editorial philosophy, direction, and content of the journal during his or her tenure as EiC and sometimes also for the overall design of the pages. Although some academic Editors may get a small stipend for their editorial work, or perhaps get a reduced teaching load, many get no compensation for the journal work they do. In addition to leading the journal direction and philosophy, an EiC may look at all the submitted manuscripts and decide which should be rejected without review, may choose the Associate Editor who will oversee the passage of a particular paper through the peer-review process, and may make the final decisions and write the decision letters. Alternatively, the EiC may delegate more of these responsibilities to an Associate Editor, sometimes called a Subject Matter Editor, or to a paid staff member, often called a Managing Editor, Senior Editor, or Assistant Editor. The EiC is usually responsible for appointing the Associate Editors or Subject Matter Editors to the editorial board and probably makes final decisions about the hiring and firing of the paid staff. The Editor-in-Chief is also almost invariably the final arbiter when it comes to appeals.

Associate Editors or Subject Matter Editors

As mentioned above, the Associate Editors (or Subject Matter Editors) are usually volunteers—scientists or other specialists who form the editorial board. These individuals are chosen for their in-depth knowledge of a particular topic within the overall subject area that the journal covers. The bigger journals, and those with multiple issues per year, require a greater number of Associate Editors to deal with the large number of papers submitted for publication. They will be approached by the Editor-in-Chief or by journal staff and invited to oversee the peer review process for those papers that fall within their area of expertise. Once they have read the manuscript, they are often required to pick relevant peer reviewers, since they are likely to know most of the other experts in their field. The Associate Editor receives the peer reviewers' reports and either writes the decision letter or writes an overall recommendation that the Editor-in-Chief (or a journal staff member) then relays to the author.

The editorial board

Some journals are run by an editorial board, headed by an appointed chairperson, rather than by any one individual. In such cases, the day-to-day running of the journal is usually in the hands of a paid staff member, whose title is often Executive Editor or Managing Editor (see below).

The editorial board meets at regular intervals, usually once or twice a year. The chairperson, who may be the Editor-in-Chief or someone chosen from among the Associate Editors, sets the agenda and runs the meetings while someone else, often a staff person, takes minutes. These board meetings are an opportunity for the Associate Editors to meet their fellow board members, discuss new plans for the journal, exchange views, give general feedback to the staff, and make decisions about future content.

STAFF POSITIONS

Executive Editor or Managing Editor

In journals where the EiC is a practicing scientist, based in a different institution or even in a different city from the editorial offices, there will often be a member of staff with the title Executive Editor or Managing Editor. This is a very senior position, held by a highly experienced publishing professional, who answers directly to the EiC, or to the chair of the editorial board, and who is in charge of other journal staff and of the day-to-day running of the journal, and usually has responsibilities in connection with the peer review process as well. Executive Editors are likely to have been scientists themselves at an earlier stage in their career, though not necessarily in the same field. Having made the decision to switch from science to publishing, they will probably have started at a junior staff level and worked their way up through the ranks, becoming familiar with all parts of the publishing process along the way.

Assistant Editor or Senior Editor

In the offices of a large weekly or monthly journal, a number of Assistant Editors will be administering the peer review system and keeping manuscripts moving through the publication process, from submission and peer review, through copyediting and proof checking, to final publication. They may also play a role in the decision-making process. If you have a general question about the journal, or about the progress of your manuscript, an Assistant Editor is the best person to contact. Assistant Editors may also be involved in the editing and preparation of your manuscript once it has been accepted, so if questions arise in the journal offices regarding your paper, it is probably an Assistant Editor who will contact you for information.

Copyeditor

Where present, usually only in the larger journal offices, Copyeditors edit papers to improve grammar, punctuation, and clarity, and to ensure that the text conforms to journal style. They are not usually involved in peer review or the decision-making process.

Editorial Assistant

Editorial Assistants are the most junior staff members in the team. They may do administrative work on the journal, and possibly some copyediting and proofreading. Occasionally, especially on smaller publications, it is the Editorial Assistant who administers the peer review system, though they will not be involved in any decision making.

THE BOTTOM LINE

Peer review ensures that the information published in scientific journals has been examined by a number of experts in the field who have judged it to be scientifically sound. All the current models of peer review have advantages and disadvantages, and it is advisable to have a good understanding of these. Having your papers peer reviewed, and performing this service for others, is an integral part of most scientific careers and will help to improve both your critical thinking and your writing skills. When you are a reviewer, you should perform the kind of thorough, constructive, and timely review that you hope someone would do for you.

CHAPTER TEN

Dealing with decision letters

A rejection is nothing more than a necessary step in the pursuit of success.

BO BENNETT, businessman

One of the hardest parts of the publication process for many authors is the period between submitting the manuscript and the arrival of the decision letter. In the past, of course, it actually was a letter, so that the waiting time was increased by days or even weeks of progress through the postal system. Now, the Editor's decision is usually delivered in the form of an email, which can travel from one side of the world to the other in less time than it takes to actually read it.

Different journals will vary in terms of timing and details, but the decision letter will deliver one of the following verdicts:

- rejected
- major revision
- minor revision
- provisionally accepted
- accepted.

It is vital that you read the decision letter carefully and interpret it correctly. Here, we review each category and try to help you to consider your options in each case. If you are the sole author, you will undoubtedly get advice and suggestions from supervisors and colleagues but the final decisions will be up to you. However, where you are one of many authors, the whole group will need to reach a consensus—if you are the corresponding author (see chapter 4 for details on the roles and responsibilities of corresponding authors) it will probably be your task to gather everyone's ideas and summarize them, to help the group in coming to an agreement.

REJECTED WITHOUT REVIEW

We have already discussed the rejected-without-review category in earlier chapters and described ways to avoid this highly undesirable outcome. However, since every author's primary objective is to get past this hurdle, we discuss the topic briefly again in this context.

If you receive a short letter, telling you the paper has been rejected without any form of peer review, and with no other explanation, you can contact the Editor and politely ask for more information. Many, though not all, Editors will respond to such a request, particularly if it comes from a young or inexperienced author.

The Editor will very probably reject your manuscript without review for any of the following reasons:

- You sent the paper to the wrong journal (wrong topic, wrong level of writing, or wrong type of paper).
- The journal publishes only cutting-edge, high-impact science and has a high rejection rate (you aimed too high).
- The journal has a big backlog or a lot of papers on a similar topic and does not wish to consider any more (that is just bad luck).
- You didn't explain your ideas clearly enough. (Are you a nonnative speaker of English? Was there a language problem? If so, consider getting help.)
- Your science is flawed (see below).

If the letter explains why the paper didn't make it past the Editor's desk, then at least you've gained some useful information. Was the problem in the scientific content or was this rejection based on a mismatch between manuscript and journal? Did the Editor spot a flaw or omission in the methods section or were you just a bit overambitious in sending a relatively minor advance in understanding to quite such a prestigious journal?

You need to read the letter carefully and then go back to basics. Follow any advice in the letter regarding how to improve or reformat the paper and then restart the process of choosing the right journal to send it to (see chapter 5, "Choosing the Right Journal"). If the Editor has "left the door open" for you and is inviting you to submit again, once you've made the requested changes, then your course of action is clear. If the Editor has suggested one or more alternative journals to submit to, you should study these carefully; it is your responsibility to make sure that these publications really are a good match for your paper before following this advice.

REJECTED FOLLOWING REVIEW

The manuscript has been sent to reviewers and the Editor has received one, two, or three reports. The reports may be from two reviewers, together with an overall recommendation from an Associate Editor or, if the first two reviewers had opposing views, a third reviewer may have been called in to act as a tie breaker. In either case, the overall recommendation was rejection (see chapter 9, "Who Does What in Peer Review").

Whether you have received a decision of either "rejected following review" or "major revision," you are going to need to use the reviewers' suggestions and comments to improve the paper before submitting it again. Whether you send your revised paper back to the same journal or submit it somewhere else, you need to take a similar approach in responding to the reviewers' comments. Here are a few important points to remember, as you begin the painful process of overhauling your paper.

First, when studying the reviewers' reports, don't take offense. Reviewers (and Editors) are not always as tactful as they could be, and some irritation or impatience does occasionally creep into their comments. You should ignore these remarks; instead, study the criticisms as unemotionally as possible and take what is useful from them. No matter what tone is adopted, you need to try to view the information as valuable assistance with your paper because, difficult as it may be to accept, most reviewers' comments *are* valuable and can help you to improve the manuscript. Also, if the paper was rejected, make sure you understand why. As mentioned earlier, there can be a number of reasons for rejection, and some are easier to fix than others.

Here are some common problems that will trigger a rejection letter.

The subject matter (or the format) of the paper does not fit the journal

If this is the problem, then you didn't do your homework properly. Go back and study the instructions to authors and copies of the journal again. Did you misinterpret the information provided? Why did you think your paper was a good match for this particular publication? Were the instructions unclear or not extensive enough? Having said all that, if your paper was rejected for any of these reasons after it had been reviewed, the Editor may not have been paying enough attention either. The problem should have been picked up during the initial examination of the manuscript.

There is little room to argue with a verdict like this. Start the journal search procedure again and this time, make sure that the paper really is a good match for the next journal you submit to. Start researching possible target journals before you begin making any corrections, so you can change the paper to fit the new journal's requirements at the same time as you are improving the content.

The paper is scientifically flawed

If the reviewers find your paper to be scientifically or logically flawed, your decision letter should include a detailed list of all the things that are wrong with it, thereby giving you the information you need to set about systematically addressing the problems so that you can resubmit elsewhere. Table 10.1 lists some of the more common reasons for this category of rejection.

Of course, making the required changes is not always possible. If your research was carried out during a never-to-be-repeated field trip to Outer Mongolia and you failed to collect a whole category of data, you are out of luck. If you omitted to take a key measurement from the 15,000 patients in your study, again, there is no going back. If, on the other hand, it is a question of carrying out different statistical analyses or adding a section that you failed to include in the discussion, then all is not lost. You should read the comments in the letter carefully to make sure you understand what the reviewers saw as flaws and what they are suggesting you do to address the problems. They may be asking you to:

- correct errors
- add more data
- add new analyses of the data
- change your interpretation of the results
- provide more explanation
- add or delete one or more sections
- change the conclusions or provide a more balanced view.

You need to work through each set of comments systematically and make sure that your final responses to comments are organized in a logical and straightforward way that will be easy for Editors to follow. These lists of suggestions can sometimes be contradictory or confusing, so it may be helpful to group the various points from each report into categories (see later for suggestions on how to do this). Decide which points (from different reviewers) are saying roughly the same thing and which appear to be sending you in opposite directions. If you really get stuck and your supervisor or other colleagues can't advise you, you may be able to appeal to the journal for

TABLE 10.1. Common reasons papers are rejected following review

Category	*Translation*
Science is not novel.	Something similar (or identical) has already been published.
Not enough new information.	The science does not represent enough of an advance to warrant publication.
Arguments are unsound.	Data are poorly analyzed or incorrectly interpreted.
Arguments are weak or the evidence presented is unconvincing.	Conclusions made on the basis of these data are unsupported or the importance of the data is exaggerated.
Data or other evidence are missing.	Certain experiments were not done or previously published findings were not taken into account.
Important arguments were not addressed or seminal work was ignored or not cited properly.	Again, some findings (your own or that of other researchers) have not been taken into account in building your argument.
Paper is biased.	In a scientific question with two or more possible hypotheses or interpretations, the paper highlights one side of the argument, rather than presenting a balanced view.
Arguments are unclear and confusing.	The research or conclusions may be valid but are not clearly explained; alternatively, there may be something wrong with the way the basic premise, or the approach, or the results, or the conclusions have been presented.
Paper is uninteresting.	The Editor may feel that the paper is boring, or will not enhance the journal, or will not be of interest to readers. Unfortunately, papers describing negative results may be included in this category, even though the resulting publication bias is a major problem in scientific publishing.
The paper was not prepared according to the guidelines set out in the instructions to authors.	If the manuscript clearly doesn't fit the criteria listed in the instructions, and this fact was missed at initial review, it may mistakenly be sent to reviewers. If the Editor feels that it would be impossible to change the paper sufficiently to conform to the journal criteria, it will be rejected.

help. Write to the Editor who signed your decision letter, explaining your predicament, and ask them which reviewer's recommendations they would prefer to see carried out (see "What to do when peer reviewers contradict each other," later). In cases where the suggestions are ambiguous or otherwise difficult to understand, and where the reviewers are anonymous, contact the Editor and ask (a) if you can be put in touch with one or both reviewers, so you can discuss the issues with them directly, or (b) whether you can provide a list of questions for the Editor to pass on, so that the peer reviewers can respond while still maintaining their anonymity.

You should also consider the possibility that the problem may lie not so much with the science itself but with the way you explained it. If this is the case, you need to redraft the text, following the reviewers' suggestions, and then ask colleagues (preferably people who are not too familiar with your work) and friends to read it, to make sure your descriptions and explanations come across clearly.

Once the paper has been revised to the best of your ability, the next question is where to submit—the same journal or another publication? As mentioned earlier, the decision letter may indicate that the Editor is willing to consider the paper again, if you can change the manuscript according to the reviewers' suggestions. However, if the letter says something like, "We hope you will find these comments helpful in preparing the manuscript for submission to another journal," take the hint. The Editor does not want to see your paper again. Unless you feel you have very strong grounds, don't bother to appeal.

At the end of the letter, you may find that the Editor has suggested another journal to submit to. As mentioned above, you should not just blindly follow this advice. The Editor may have made a useful suggestion or may just have mentioned the first title that came to mind, without stopping to think whether it really is a suitable venue. It is your responsibility to investigate the suggested journal to make sure you are not setting yourself up for another rejection letter.

The journal has too many papers on this topic already

Space is always at a premium in print journals, although less so in online-only titles. Nevertheless, even in electronic publications the Editor must maintain a balance between all the different topics within the specialty. Very few journals would want to publish several papers on much the same aspect of their discipline, so if several such papers are submitted at roughly the same time, the Editor will choose a small number to review and reject the rest. In a case like this, there may be nothing wrong with your paper; it just arrived in the right place at the wrong time. All you can do is move on and choose another journal.

A RAY OF HOPE: REVISE AND RESUBMIT

Different journals have different cutoff points between "rejected" and "major revision." A publication with a high submission rate or a serious backlog of articles is much more likely to reject borderline papers rather than to ask for major revisions. However, if your paper is on a topic the Editor is particularly interested in, or if the basic premise of the manuscript is good but the execution is poor (see table 10.1), the Editor may be prepared to consider the manuscript again after you've made the suggested revisions. If you get a letter inviting you to revise and resubmit, remember that this is no guarantee that the paper will be accepted the next time around. It does mean that you still have a chance of publishing a version of the paper in that particular journal and that you don't have to carry out another journal search right away. When writing the cover letter to accompany the revised version of the paper, be sure to remind the Editor that you were invited to resubmit. Editors deal with hundreds of papers and cannot be expected to remember every one, so your reminder may be what tips the balance and gets the paper sent out to peer review again.

If the paper is rejected and you are invited to resubmit, the revised version of the paper will be counted as a new submission and will probably have to go through the entire peer review process again. It may or may not be assigned to the same Associate Editor and it may or may not be sent out to the same reviewers. This may or may not be good news. The previous Associate Editor and reviewers clearly didn't like the paper (or they would not have recommended rejection), so perhaps a fresh set of eyes will view it in a more positive light. However, all you have to go on as you try to improve the manuscript are the comments made by that first set of reviewers. Even if you submit the paper that they have basically "advised" you to write, you have no guarantee that the second "judge and jury" will agree. Indeed, they may come up with an entirely different set of criticisms and another recommendation to reject.

MORE TIPS ON RESPONDING TO DECISION LETTERS

Rejection

Rejection letters can cause a whole range of emotions from mild disappointment and annoyance to deep distress or red hot rage. Nevertheless, no matter how idiotic, blind, or downright incompetent you may feel the Editor and reviewers to be, you must never ever give in to the temptation to tell them so. You may be meeting them at a conference some day, they may well be colleagues you already know or have corresponded with, or, worse still, they may be on a selection panel for a job you are applying for in the future. In most cases, you will not know who your reviewers are (since many journals still employ a single-blind reviewing system), but they will know who you are. Despite the brief moment of satisfaction you might get from writing a sarcastic or overly frank response, you may have to pay a heavy price in the future.

If you are going to respond at all, make certain that your letter is polite and professional and lays out your arguments in a calm and reasonable manner. A thoughtful and logical response will do no harm and it may actually help persuade the Editor that you are right and the reviewer wrong (see "The appeals process," later).

Major revision

If you get a letter asking for major revisions, you need to focus on the fact that although there is still a lot of work to be done, there is the possibility of a publishable paper in your submission. You have the reviewers' reports and can get to work. Remember, however, that a verdict of "major revision" is no guarantee of acceptance. If you don't do a good job with the revisions, the paper may still be rejected.

When revising your paper, you need to make the suggested changes as soon as you can. Most journals have a cutoff date, which may be three or six months, or occasionally longer, after which the manuscript will usually be marked as "withdrawn" and the file will be closed. If you send in the manuscript after the cutoff date, it will be treated as a new submission and you may have lost an important advantage, namely the opportunity to remind the Associate Editor and reviewers that they had seen the possibility of a good paper in the first submission. Getting the paper back in a timely manner will often ensure that the second version is seen by the same reviewers, if they are still available, which is what you want since the revisions you made were the ones *they* asked for.

Dealing with long lists of comments

Detailed peer reviewers' reports are very helpful, since they will hopefully guide you, step by step, towards a better paper. However, very long, detailed lists of comments can be rather daunting, especially when you have received two or even three sets. You may wonder how you are ever going to make sense of all those suggestions and criticisms, let alone apply them sensibly to your paper. "Divide and conquer" is the best strategy for dealing with this kind of information overload. You can organize the lists of recommendations in a couple of different ways to make them more manageable, and both involve dividing and rearranging the comments into separate groups.

The first strategy is to prioritize. Pick out all the major suggested changes from each list that the different reviewers have provided and put them together. Make a further list of second-level suggestions, and then a third list of small changes, involving spelling, word choices, and so on. Create each of these categories as separate blocks, with space between each comment. The easiest way to do this is to copy all the comments into a blank document and then make them all bold or italic, or a color other than black. Separate each comment by a few line spaces, and then, in a normal, unbolded, black font describe the changes you made in response to each point. Tackle the big changes first, addressing one recommendation at a time and applying it to the paper. After each change has been made, write out what you did, together with any comments, questions, or problems you encountered—this way, you are building up the basis for your response letter (sometimes also called a rebuttal letter) as you go. Link each comment to the relevant line numbers, so the Editor or future reviewers can find each change easily. Once you have dealt with the more complex suggested changes, work on the second-tier list, again keeping a running commentary of what you have done and what you could not do, for whatever reason. Make the small corrections at the very end; many of these may have been dealt with already, as part of the bigger suggested revisions.

Another strategy for dealing with long lists of reviewer suggestions is to divide the comments according to the sections of the paper they deal with. Again, assemble comments from all the different lists you received, and group them according to the section they pertain to, for example the introduction, methods, results, discussion, or conclusions. Then work through the comments for each section, one at a time, creating your response letter as described above. Of course, if the paper is a review, this method may be harder to carry out, as such papers are not always so easy to divide into sections.

Even if the journal does not explicitly request this, you should always accompany the second and any further submissions of the manuscript with a point-by-point explanation of exactly what you did in response to each of the reviewers' comments. The Editor and reviewers will find it much easier to check the manuscript if they have a list of your responses to their comments, as this will save them time as they search the paper for the various revisions they requested.

Work through the lists systematically, crossing off each item, or changing the color of the type, as soon as you have dealt with it, so that every comment is addressed and you are not in danger of missing any. If the decision was for major revision, then the resulting response letter will go back to the Editor, together with the new draft of the paper. If the decision was to reject, then you should still use the reviewers' comments to improve the paper before sending it to another journal.

Reviewers' comments you disagree with

Within the list of comments, suggestions, and criticisms from the reviewers, there will probably be a few you just don't agree with. As you write your point-by-point responses, it is acceptable to say that you disagree with the reviewers on a couple of issues, provided you can give a good explanation as to why you believe those particular comments to be wrong. After all, it is your name at the top of the paper—you must believe in what you have written and should be able to justify and support every point you make. However, you cannot argue with the reviewers too often. If you find yourself disagreeing with more comments than you agree with, then you have a problem. The Editor is likely to feel that you are being defensive and argumentative if every second response to a comment begins, "I don't agree with this point because . . ."

If the issue is one of interpretation—in other words, you think the reviewers misunderstood what you were trying to say—consider how you might revise your text to make your meaning clearer. Can you add further evidence? Can you frame your arguments in a different way? Can you find a compromise that will satisfy both sides? One option is to describe the two possible interpretations (yours and the reviewer's) and explain why you believe your view is the right one. Finally, you can write a polite email to the Editor, explaining the situation, if you really disagree with a lot of what the reviewers have said and cannot bring yourself to follow their suggestions. You could also consider withdrawing the paper so you can submit it elsewhere and hope for a different set of reviewers.

What to do when peer reviewers contradict each other

Authors become very confused when the comments of the two peer reviewers contradict each other. Whose guidance should be followed? In fact, you should not be faced with the dilemma of choosing between two contradictory pieces of advice. The Associate Editor or Editor-in-Chief should have spotted the problem while reading all the reports, prior to writing the decision letter. They should then have provided some guidance in the decision letter on what you should do. However, in reality, busy Editors don't always have the time to read all the peer reviewers' comments that carefully and so may have missed the contradiction, or they may have decided to just let you make your own decision. In either case, if you really don't know which way to go, contact the Editor who signed your decision letter. You may get the guidance you need to navigate around this particular problem. If you agree strongly with one reviewer rather than the other, and you can make a good case for taking your paper in that direction, then you should explain your decision carefully in your response letter.

Minor revision

If the Editor and peer reviewers feel that only minor changes are needed to make the paper acceptable for publication, then your decision letter will tell you so. Minor revisions should be the sort of changes you can easily do within a couple of weeks to a month, depending on how busy you are. If you are about to leave to do fieldwork for any length of time, or you are going on vacation, and you won't be able to supply a revised version of the manuscript reasonably quickly, then write and tell the Editor so. That way, they will not assume that the paper has been withdrawn and close the file if the changes don't come in promptly.

Even for minor revisions, you still need to supply your list of point-by-point responses. Again, this list will save the Editor and reviewers from having to search the manuscript for the changes you made. Reviewers may have suggested lots of small corrections (some reviewers will highlight every typographical error and stray comma) or may list just a few substantive but relatively straightforward changes. In either case, work your way through the list, add any extra comments or questions in an accompanying cover letter, and resubmit as quickly as you can. The chances are high that your next letter from the Editor will inform you that the paper has been accepted.

If you are submitting to a print journal, you may be required to submit the high-resolution versions of all your figures at this stage, if you have not already done so (see chapter 8 for more information on submitting figures) and there may be other issues or documents to deal with, depending on the journal. Read the Editor's letter very carefully and make sure you have done everything that has been asked of you.

Provisional acceptance

Not all journals use this category. If your manuscript is provisionally accepted, it basically means that the manuscript is almost ready for acceptance and publication, but you still have a couple of final matters to attend to. You may need to do something as simple as supplying high-resolution versions of the figures or signing a form that you've forgotten to return, such as a copyright agreement or author billing form (in cases where page charges or an open access publication fee is required). You are nearly at the finish line with this paper. Do everything you have been asked to do and then relax—that longed-for acceptance letter is sure to follow shortly.

ACCEPTANCE AND AFTERWARDS

Even once the paper has been accepted, there are still important tasks to perform. Some of these may be listed in the letter of acceptance, so don't stop reading once you have got to the bit that says, "I am delighted to be able to tell you. . . ."

Unless the letter specifies this information, it is a good idea to find out roughly when the paper is scheduled to be published or, more importantly, when you might expect to receive page proofs or an edited version of the manuscript for checking. Make sure that you and your coauthors will be available at that time. If there is a lengthy field trip or sabbatical coming up, inform the Editor of the dates when you will be away (some manuscript submission systems allow you to log times you'll be unavailable). To avoid any unnecessary delay, you need to make sure that all your coauthors have signed copyright transfer forms, that you (or the relevant person) has filled in the author billing form and the order form for either the reprints or a pdf, if these are available for purchase, and that any other paperwork has been dealt with.

Once your paper has been formally accepted, you may want to let the Editor know if there are any specific reasons why it should be published quickly. There is no guarantee that the Editor can or will grant your request, but if the reason is truly important your paper may, in some cases, get pushed towards the front of the line. However, some journals specialize in rapid publication, so if speed is an important consideration, you may want to consider submitting to this type of journal.

Proof checking

In the past, authors were sent a paper proof of their manuscript so they could check that editing had not changed their meaning and that no errors had been missed or introduced. This was often a long scroll of paper, known as a galley proof, which showed the text in one long, continuous column. These days, although the word "galley" is sometimes still used, the proof you receive is most likely to be an electronic copy of the paper, laid out in the format in which it will appear in the journal. Usually, this will be accompanied by a series of questions, either as a separate document or as a series of notes in the margins of the proof. If you are the corresponding author, send the proof and questions to all your coauthors immediately, so they can check everything at the same time as you do. They should send all their corrections back to you, as it will be your job to assemble them all onto one proof and return it to the journal offices (see chapter 4 for more information on the duties of corresponding authors).

Read through the text and questions extremely carefully and only make corrections that are absolutely necessary, since some journals will charge you for author corrections. This is not the time to make aesthetic adjustments to grammar or syntax—the journal staff should have done that for you. You also need to resist the temptation to change everything back to the way it was before. No author likes to see their writing altered, but, equally, an experienced Editor will improve the flow of the text and correct errors that you missed.

Dealing with the media and embargoes

Another group of potential pitfalls that await the novice author of an accepted paper involve dealing with the media, be it newspapers, magazines, radio, or TV. Many journals have strict rules about media coverage before the paper has actually gone into print or been posted online and you need to follow these, or else, in the case of the really top-tier journals, risk having your paper withdrawn from publication. Admittedly, this is a worst-case scenario and in fact most research papers do not warrant a media frenzy—most are of no interest to the media since they are of no immediate interest to the general public. Nevertheless, an understanding of the term "embargo date" and the rules that surround it is a must, in case the day comes when you are about to publish a really hot piece of research and the press are clamoring to write about it.

The word "embargo" has a number of different meanings, but in the publishing context an "embargo date" is a specific day and time, set by the journal or publisher for a specific paper; no media coverage is allowed until that day and time, when the embargo is said to have been "lifted." What this means is that no media outlet (e.g., a newspaper, another scientific publication, or a TV station) can actually tell the public about your work until that date and time. Most journalists and other media professionals are very aware of the rules surrounding embargo dates, but in the case of a really important story someone may be strongly tempted to "scoop" their rivals and publish the news before everyone else.

So, what does this mean for you, the author of a paper that is of interest to the media? Once a paper has been officially published, you can talk to whoever you like. *Before* the paper is published, you need to find out from your Editor or publisher whether there is an embargo policy and, if so, what the embargo date is for your paper. Here is what you can and cannot do before that date:

- You *can* give talks at conferences, seminars, and other scientific gatherings.
- You *can* hand out draft copies of the paper to colleagues, and discuss your findings with them, but it is highly advisable to mark every page of the paper with the words "Embargoed until DATE" in case any reporters get hold of a copy.
- You *can't* talk to individual reporters or other media professionals or take part in press conferences unless you have specific permission from your Editor or publisher.
- You *can't* provide reporters or other media professionals with copies of the manuscript, again unless you receive permission from the Editor *and* they are clearly marked with the embargo date.

If your paper really is likely to be of interest to the media (and therefore to the general public), the journal may choose to do a press release; your own institution may also do one. Normally, the Press Officers of both the journal and your institution will liaise with each other and with you before distributing a press release. They may give out copies of the paper, stamped with the embargo date, to journalists on their regular contacts list, and may allow you to be interviewed, so that news stories can be written, ready for the moment the embargo is lifted. At that time, the stories will be posted online or will appear in the newspaper or journal they are writing for. If you really are about to publish a newsworthy piece of science, contact the Editor of the journal and the Press Officer of your institution to discuss this. Find out the embargo date and make sure you will be available for interview at that time. Get advice from the Press Officers or from seasoned, media-savvy colleagues about what to say and how to say it when being interviewed. Talking to the press can be tricky, so be prepared: work out how to explain your science in straightforward, everyday language. Don't use specialist terminology and don't expect the interviewer to follow complex explanations. Translate your science into an easy-to-understand story and try it out on nonscientist friends or relatives. Did they understand? Did they find it interesting?

Many journalists are friendly and easy to talk to, but a few can be a bit aggressive in their interviewing style, particularly if the story is going to be controversial. Stay calm, answer clearly, and don't allow yourself to be rushed or pushed into saying something you don't know for certain.

Dealing with the media is not really within the scope of this book, so the above section has just touched the surface of this topic. We strongly recommend that you to get advice and help before you find yourself at the other end of a microphone, being peppered with questions.

NEW EDITING TECHNOLOGY

The paragraphs above relate to journals that use editorial staff to prepare the paper for publication. However, in order to streamline and speed up publication following acceptance, some journals now process papers through special editing software that can identify problems such as spelling, punctuation, and grammatical errors, or discrepancies between the citations in the text and those in the References list. These programs will also alert journal staff if you have mentioned a table or figure but have not provided it, which is a fairly common error, but they cannot clarify badly explained ideas or pick up a mistake in your calculations, in the way that an Editor often will. Whether you are publishing in a journal that uses editing software or in one that employs editorial staff, you should always check your text with extra care before sending it in.

THE VERSION OF RECORD

Errors in a published paper are embarrassing to the author and a source of extreme annoyance to Editors, since no Editor likes publishing errata (the notices that journals publish when a substantive error has been found in a previously published paper). Authors often wonder, in these days of online-only publishing, why editorial staff cannot simply make the necessary changes online, as soon as mistakes are identified. Editors have wrestled with this topic ever since electronic publication became possible. After all, when something is printed on paper, there is no way to change it, but making changes online is easy, right?

Wrong! What you need to remember is that the version of the paper that is first published, be it in print or online, is the "version of record" (see also sidebar 3.2 on CrossRef DOIs, CrossCheck, and CrossMark). As soon as it is available to view, people will start reading it. If authors later make substantial changes to the online paper, then future readers will be seeing a different version of the text and it would be impossible to tell who had seen which version. So Editors have developed various ways of indicating that an error has been found and that a change is necessary. In an electronic journal, this may be in the form of a "sticky note" next to the relevant section of the text, while in a print journal it will appear as an erratum in a subsequent issue.

In short, avoid the embarrassment of having to ask the journal to publish an erratum about your paper by checking, rechecking, and then checking again to make sure that every aspect of the paper is correct before it goes to press.

THE APPEALS PROCESS

As mentioned earlier, you should never allow your emotions to influence your response to a rejection letter. Usually, a little time and the immediate ingestion of chocolate or alcohol will sooth the pain and you will start to see what you can do to get the manuscript back on track. Do nothing for at least twenty-four hours or until any strong emotions have subsided. In most cases, once you have calmed down and consulted with supervisors, colleagues, and friends, you will be able to set about improving the paper and continuing your attempts to get it published. Very occasionally, you may have to give a particular manuscript up as a lost cause and move on to something else.

In very rare cases, however, having looked at the paper and the reviewers' comments from every angle, and after discussing the situation with your advisors and colleagues, you may feel strongly that you have cause to appeal the decision. In such situations, you can write a polite letter to the Editor-in-Chief, or the appropriate Associate Editor. Taking each point in turn, lay out your reasons for requesting a re-evaluation of the manuscript. Explain which points you agree with and can fix, and why you believe the reviewers have misunderstood your reasoning in some places, or are just plain wrong. If your arguments are strong, the Editor may reverse the decision and send the paper out for further review.

Do not try to be clever, by guessing and naming the writer of an anonymous review (e.g., "Professor Black has never liked my work and is trying to get his own paper, on a similar topic, published first"). Whether your guess is correct or not, this will do nothing to further your cause. Clear, rational reasoning is your best weapon. Make a convincing argument and you just might get the Editor to reconsider the decision, giving you a chance to resubmit or sending the manuscript back out for further review.

REJECTED—NO APPEAL—NOW WHAT?

Before you even begin to plan your next move, you should take a moment to remember this: every author gets rejected at some time, up to and including

Einstein, so you are in good company. Younger, less experienced authors tend to get rejected more frequently, which is unfortunate since they are at a stage in their career when they desperately need publications and when rejection can do the most damage to their self-confidence.

If you get rejected and have no grounds for appeal, you need to take an objective look at your paper and decide on the best way forward. First, get advice. Discuss the paper with supervisors or senior colleagues and get their ideas on where to resubmit. If the Editor hasn't made any suggestions, you can write and ask. You may not receive a reply, but it is worth a try. Whatever you do, don't just take the manuscript and send it straight off to the next journal. Carefully research the new publication's requirements and change your paper and cover letter accordingly.

Alternatively, you might try thinking outside the box. Consider taking the paper apart to make something new: for example, you could develop the literature review in the introduction into a review paper, or split out the methods and submit as a methods paper, or focus on one important finding and format it into a short research communication or research letter. And, of course, you can also use the peer reviewers' and Editor's comments to strengthen and improve the manuscript.

Do remember that a new Editor or a new set of peer reviewers may very well contradict everything the previous ones said, or may ask for a completely different set of changes. Authors who have carefully followed a long and complex set of recommendations from one set of reviewers can get very frustrated when asked to reverse everything they have done by the next journal they submit to. Conversely, it is perfectly possible for the new target journal to send your paper to the same reviewers as the previous journal did. In that case, those reviewers will be looking to see whether you followed all the suggestions they made before, and may be annoyed if you ignored a large number of their comments. In short, there are no guarantees that the changes you make will be the right ones. Create the strongest paper you can and see what happens. Nobody said that publishing scientific papers was easy.

THE BOTTOM LINE

Decision letters may bring good news or bad, a lot of helpful information or, more rarely, almost nothing of any use at all. Read the letter carefully and remember that most decision letters contain the roadmap to a better paper and eventual publication!

CHAPTER ELEVEN

Ethical issues in publishing

Principles:
Honesty in all aspects of research
Accountability in the conduct of research
Professional courtesy and fairness in working with others
Good stewardship of research on behalf of others

Singapore Statement on Research Integrity, July 2010

In this chapter we will briefly examine various ethical issues that commonly arise in scientific publishing. Importantly, we look on both sides of the fence, considering unethical practices not only by authors but also those committed by Editors and peer reviewers. As an author, you will need to be aware of these issues, so that you do not infringe the rules yourself, and so that you have a better chance of recognizing when an Editor or peer reviewer associated with your paper is behaving in an unethical manner.

Luckily, serious unethical conduct in both authors and Editors is relatively uncommon, so that when a case of major misconduct does come to light it tends to make headline news, at least within the science press and, increasingly, in the local or national press as well. More frequently, researchers engage in small-scale cheating, sometimes deliberately, sometimes not. This more subtle kind of unethical behavior can be hard to detect and difficult to deal with. Editors who discover such infractions have to decide just how severe they want to be with an inexperienced researcher who accidentally or deliberately used a piece of copied text or a figure from another published paper in their own work. And what can an Editor do who suspects that a peer reviewer may be biased against an author (knowing, for instance, that these two individuals had a serious disagreement in the past), other than not asking that individual to review for the journal again?

Ethical dilemmas in publishing are arising with increasing frequency; entire books have been written about evaluating and responding to these circumstances. Although this short chapter cannot do the topic justice, we nonetheless discuss some of these issues in brief, to underscore that authors and Editors alike must be attentive to potential ethical conflicts and unethical behaviors, and remain vigilant in attempting to prevent such behavior in any realm of scientific publishing.

UNETHICAL BEHAVIOR BY AUTHORS

Unethical behavior by authors can be divided into two subcategories—unethical behavior while conducting research and unethical behavior in trying to get it published. It is outside the scope of this book to go into detail about unethical practices in science, especially since there are many publications devoted to the subject (see the list of recommended readings at the end of this chapter, and sidebar 11.1 for a discussion of this topic).

SIDEBAR 11.1

Ethics in scientific publishing

DIANE SULLENBERGER
Executive Editor, *Proceedings of the National Academy of Sciences of the United States of America* (PNAS)

When manuscripts are submitted for publication, Editors expect the research to have been conducted in compliance with high ethical standards that contribute to the advancement of scientific research and to be worthy of the public's trust in the scientific enterprise. If either submitted or published work is found to be in breach of these ethical standards, the consequences for the researchers may include any of the following: a written warning, an institutional misconduct investigation, a ban on submitting to the journal for a certain number of years, a published notice of "expression of concern" about a work, a published notice that the work has been retracted, loss of research funding, termination of employment, or—in the case of Eric Poehlman, who was convicted in the US of falsifying data to obtain federal grants, a jail sentence (Interlandi 2006). To steer clear of such unfortunate consequences, authors should be diligent in learning what responsible conduct in scientific publishing is, and must then adhere to those principles.

Research misconduct, as defined by the US Department of Health and Human Services' Office of Research Integrity, means "fabrication, falsification, or plagiarism in proposing, performing, or reviewing research, or in reporting research results" (http://ori.hhs.gov/misconduct/definition_misconduct.shtml). Fabrication means "making up data or results." Falsification means manipulating the research in order to portray it inaccurately. Unethical practices such as fabrication and falsification consume tremendous resources when other researchers try to replicate the work; it also squanders research funds, misleads the public, and may have public health or other serious consequences. Researchers should always truthfully represent their research and their findings.

Plagiarism involves using the words, data, or ideas of others without attribution. Plagiarism can ruin an author's credibility and basically means stealing from those whose works have been presented as the author's own. Some Editors use electronic plagiarism-detection systems like CrossCheck (see sidebar 3.2) to screen articles before publication. A simple rule to avoid plagiarism is to always cite the source of material, even if it has been reworded.

Although the most basic principles of responsible scientific conduct are that researchers should scrupulously avoid fabrication, falsification, and plagiarism, there are many other important ethical principles. Most journals explicitly state the ethical policies that authors must adhere to in their instructions for authors. These policies may require:

- *attesting that the work is original and has not been previously published or concurrently submitted elsewhere*
- *limiting authorship to those who contributed substantially to the work*
- *fully disclosing any association that poses a conflict of interest in connection with the manuscript*
- *ensuring that digital images are free from undisclosed manipulation*
- *making unique materials, data, and associated protocols available to readers*
- *following ethical standards for human and animal experiments (e.g., ensuring that human study participants provide informed consent to the experiment)*
- *depositing clinical trial information into a clinical trials registry before patient enrollment*
- *disclosing if the manuscript reports potential dual-use research of concern (e.g., research that might aid bioterrorists).*

Failure to adhere to these and other stated journal policies, some of which vary by scientific discipline, may be considered research misconduct or, in some cases, inappropriate research practice. The Committee on Publication Ethics (COPE) flowcharts (http://publicationethics.org/files/u2/All_flowcharts.pdf) are an excellent resource and detail the decision processes many Editors face when they receive allegations of research misconduct or unethical behavior, such as duplicate or redundant publication, inappropriate authorship, or reviewer misconduct. One flowchart also describes how COPE investigates complaints against Editors.

All breaches of ethical policies, from falsification of the number of study participants to failure to share unique materials such as knockout mice strains or computer algorithms, have a broad impact in undermining the credibility of in-

dividual researchers, their institutions, their funding bodies, and the scientific community as a whole. Researchers have a duty and an obligation to uphold the highest scientific ethics, in principle and in action.

Many countries around the world now require universities and research institutions to have ethics boards in place. Researchers must submit their proposed experiments to the board (particularly those involving live animals or humans or procedures that could harm the environment) and must receive permission to carry out the research. In the US, any institution that uses animals for experimental or teaching purposes is required by law to set up an Institutional Animal Care and Use Committee (IACUC). Similarly, when the research involves human subjects, the institution must have an institutional review board (IRB), sometimes called an independent ethics committee (IEC) or ethical review board (ERB). When experimenting on animals or humans, the research protocols must first be approved by one of these review boards before the research can begin; many journals require that manuscripts describing experiments involving animals or humans must include a statement declaring that they received approval from the relevant board. If an Editor or peer reviewer believes that, for instance, experimental animals were subjected to unnecessary suffering or that human subjects were placed at unnecessary or unjustifiable risk of side effects or other harm, the authors will be required to show proof of approval by the IACUC or IRB, and to submit the text of the protocol, to show that the experiments described in the methods section of the paper were followed exactly.

When it comes to getting research published, there are a number of unethical actions that authors may be guilty of; in some cases, they may be completely unaware that they are doing anything wrong. Nevertheless, if the Editor finds out, then the paper will very likely be rejected. If it has already been published, it may be retracted. Both of these outcomes are unpleasant for all concerned, so it is worth going through some of the more common categories here, to help you avoid them:

- plagiarism and self-plagiarism
- manipulation of figures
- multiple publications
- multiple submissions
- gift authorship
- denial of authorship
- conflict of interest.

Plagiarism and self-plagiarism

The US Department of Health and Human Services' Office of Research Integrity defines "plagiarism" as "the appropriation of another person's ideas, processes, results, or words without giving appropriate credit." In science, you can describe and discuss the ideas, experiments, and conclusions of other authors, as long as you acknowledge the original author and properly cite the publication in which the work appeared. However, what you must not do is to use the same words and phrases as the original author when describing that work, unless you clearly indicate that you are doing so by using quotation marks. The copying of pieces of text from someone else's paper or book and pretending that the words are your own is a kind of intellectual theft and is not only plagiarism but also a breach of copyright law. Part of the problem is that if you are not fully aware of what plagiarism is, it is easy to do this accidentally. You read a published paper and then, when you must describe similar information, you use exactly or almost exactly the same language that the previous author used, either unconsciously or consciously. You need to be extremely careful not to do this, as it can result in all sorts of serious consequences, from rejection of your paper to dismissal from your job, and all the possibilities in between. The fact is that you can be accused of and reprimanded for plagiarism or copyright infringement whether you were aware of the problem or not.

Many authors ask how many words constitute a case of plagiarism. Most people seem to agree that between five and ten words is probably where the threshold lies, but again, it all depends on *which* words. If you are describing a common action, such as "we set up a series of transects through the forest" or "we admitted a total of 500 patients into a phase II clinical trial," the chances are that hundreds of other authors have used the same words because there are very few other ways of saying this, and other authors will have used those as well. Importantly, also, these words are not reflecting a unique idea or concept. However, if you were to write the sentence "Time is what prevents everything from happening at once" or "Reality is merely an illusion, albeit a very persistent one" without crediting Albert Einstein, then you would be plagiarizing, because he used them first and they reflect his unique thoughts and ideas.

There was a time when it was much more difficult to detect plagiarized text—in fact, it was usually the peer reviewers or the Associate Editor, chosen for their detailed knowledge of the field and its associated literature, who would recognize previously published text when reading a submitted manuscript. Having checked the existing literature to confirm their suspicion, they would then alert the Editor to a case of plagiarized (i.e., duplicated) text. Nowadays, however, software can do the same job much more quickly and easily (see sidebar 3.2 on CrossRef DOIs, CrossCheck, and CrossMark). More and more publishers are investing in this kind of software and running all submitted manuscripts through it to ensure that the text of each new manuscript is written in more or less original language. University professors are using similar software to detect plagiarism by students.

Authors who are nonnative speakers of English can have particular problems with plagiarism. First, many of these authors have not been properly informed about what constitutes plagiarism or how to avoid it. With that lack of understanding, authors for whom English is a second (or third) language are tempted to incorporate some of the same language they have read in other papers into their own work. Whether or not you are a nonnative speaker of English, you must be extremely careful not to use words or ideas taken from other people's work without acknowledging the original source.

If you need help with your English writing, there are many options to help you avoid plagiarism. You can collaborate with a colleague who writes better English than you do—preferably, that colleague should be a native English speaker. If that is not possible, look at appendix 1 for a list of resources specifically created to support English language learning for nonnative speakers of English.

Self-plagiarism, another serious problem that you must avoid, occurs when an author uses the same phrases or paragraphs that they themselves have used before, in a previous publication. Authors are often surprised to hear that they are not allowed to duplicate their *own* writings in subsequent papers. However, in addition to plagiarism issues, we are now also crossing into the realm of copyright law—the words and phrases have been published elsewhere. The author has assigned copyright to the publisher of that previous paper, so reproducing any of that text without permission is not allowed.

Some people consider self-plagiarism to be a somewhat less serious breach of ethics than plagiarism of someone else's work, and think of it more as "recycling fraud" (Dellavalle *et al.* 2007): at least the original thoughts and writings were your own to begin with. Nonetheless, self-plagiarism is unacceptable in scientific publishing. If you submit a paper that is found to contain blocks of text from a previous paper that you wrote, it will probably be rejected or returned to you with a demand that it be rewritten. One obvious reason for this is that, given the almost unmanageable torrent of journal articles, books, and reports being published these days, the very least that a reader should be able to count on is that each paper they read consists of original material, presented, as far as is possible, in original language. Readers do not expect to be reading the same text twice, unless the journal or book they are reading clearly states that the work has been reproduced and acknowledges the original source.

Self-plagiarism, also sometimes called duplicate or redundant publication, does not only refer to duplication of text. It can also relate to the republication of data. For example, an author carrying out a series of similar experiments may decide not to bother to repeat the controls for each experiment; Editors or reviewers may catch this type of plagiarism if they notice that the column containing the control data is the same in two different papers by the same author. Researchers have also attempted to republish the same figure, and tried to disguise this by either reversing or cropping it so that it looks different. Again, this is unethical and unprofessional behavior that can get an author into serious trouble.

Multiple publications: how many slices?

Another ethical issue you need to consider when deciding how to write up your research results is the scope of the paper or papers you might write—how many pieces can you divide the work into to get multiple publications from what is basically one piece of research? The practice of breaking up research data into multiple parts and publishing it as a series of similar papers is sometimes known as "salami publishing" or "salami slicing"; there is a fine dividing line between what is acceptable and what is likely to get you a bad reputation among journal Editors.

For example, suppose that you have carried out a major piece of research. During the course of this work, you substantially altered a well-known methodology, making it much faster and more accurate. Using this new methodology, you generated a large data set involving a comparison of the effects of a new type of pesticide on two different agricultural pest species of insect. You compared the two species and found significant differences in their response to the pesticide. Furthermore, the results from one species gave you an idea and you ran a secondary experiment and discovered that this particular species is more vulnerable to a biocontrol agent than the other one. Now, you could produce one major paper, incorporating all this information. Alternatively, you could do a brief paper on your new methodology and submit it to a journal that specializes in new methods. The data set from the comparison of the two species could go to a major agricultural research journal, citing the methodology paper. The small biocontrol experiment could go as a separate write-up to an agricultural journal that is particularly interested in alternatives to standard chemical pest-control methods. Finally, you could try writing a letter to the Editor, or even an editorial, talking about the importance of testing alternative pest-control methods because of the damage that pesticides do to nontarget species.

Provided all these papers are clearly different, and each paper is crafted to suit the target journal and its readership, this type of slicing is perfectly acceptable. However, if all you have is one large data set, comprised of tests of four different pesticides on the same two species, and you divided this into four pieces (one for each pesticide), and sent them either one after another to the same journal, or even to different journals, that would be salami publishing. These four pieces of information are what are known as least publishable units (LPUs). You might be able to get them published separately, but as they emerge it will become clear to Editors and readers that you have been salami slicing—a reputation you should definitely try to avoid.

Multiple submissions

Submitting the same paper to more than one journal at a time is something quite different from the previous example and is also discussed in chapter 8 ("Preparing for Manuscript Submission"). When an author sends the same paper to a number of different journals at the same time (in the hope that at least one of them will accept it for publication), the Editors and peer reviewers involved are unknowingly making duplicate efforts in processing, reviewing, and providing feedback about the submitted manuscript.

Figure manipulation

Another unethical activity that has gotten authors into serious trouble and caused papers to be rejected or retracted is the manipulation of figures to better support a hypothesis, to strengthen a particular feature, or just to make them look better than they did originally. Unfortunately, given the availability of image-creation and image-manipulation software and other such programs, it is extremely easy to "improve" figures.

Because manipulation of digital figures has become so prevalent, many journals have specific policies about what you are and are not allowed to do, and some have introduced measures to detect changes to figure files (e.g., images are examined specifically to detect such changes; Rossner 2002). If there is any suspicion that you have manipulated your figures, you will be required to submit the original data or samples on which the figure is based. Some journals are now requiring original data to be submitted along with every manuscript. You need to examine the instructions for authors of the journal you are submitting to, to see what is and is not permitted.

Gift authorship

A gift authorship is one where Researcher A approaches Researcher B (who may be a well-known, very senior scientist) and offers to put B's name on a paper, even though B has had little or nothing to do with the research. By presenting authorship on the paper as a gift, Researcher A may be trying to gain favor with B, in the hope of being employed by B in the future, or for some other reason. Offering or accepting gift authorships is considered unethical and should always be avoided. Researcher A should not offer, and if offered a gift authorship, Researcher B should politely refuse, since only those who have actually contributed substantially to the paper should be authors. (See chapter 4, "Authorship Issues").

Another form of gift authorship relates to the quite common practice of putting the name of the head of a research group or a supervisor as the last author listed on a paper. This practice (also discussed in chapter 4) is fine when the supervisor has done his or her job and actually supervised the research or the writing of the paper, or helped the authors in some other way. However, this practice is unethical when the supervisor has not been supportive or helpful enough to warrant authorship. The supervisor may still insist on being in the author lineup; when this happens, the actual authors may find it almost impossible to say no. Without question, this kind of dilemma can lead to long-standing resentments.

Denial of authorship

Most supervisors and senior scientists see it as a part of their responsibilities towards their students and junior staff members to help with their career development; this includes ensuring that the student's or staff member's name appears on papers, often in the first author position, if merited. Sadly, there have been cases where a particular researcher (or sometimes a technical specialist of some kind), usually in a subordinate position, does a large amount of work on a project and yet is denied authorship on the paper. Again, this has been the cause of arguments and friction among members of the scientific community.

Conflict of interest

Conflicts of interest (COIs) are discussed briefly in chapter 7 ("How to Write a Cover Letter"). For a more detailed discussion on this topic, visit the

Responsible Conduct of Research (RCR) website (http://ccnmtl.columbia.edu/projects/rcr/index.html); course 1 is on COIs.

The basic definition of a COI depends on the sector to which it belongs; this might be financial, legal, scientific, or personal. RCR defines a COI as "a situation in which financial or other personal considerations have the potential to compromise or bias professional judgment and objectivity." Editors worry about COIs because they may influence either the science being reported in a paper or the review of that science by other Editors or by peer reviewers.

Therefore, whether you are an author on a paper, an Associate Editor, or a peer reviewer providing a report on a manuscript, you *must* declare any and all COIs to the Editor of the journal. The Editor is responsible for deciding whether the situation is likely to affect your professional judgment in any way and whether to take any action as a result. For example, if you are an author, a potential conflict of interest might prompt the Editor to ask you to add a statement to the paper that describes the COI. Similarly, if you are a peer reviewer or Associate Editor, you may be asked to withdraw from that position for that particular paper if there is a potential problem. COIs can come in all shapes and flavors, but two of the most common types are those involving money and those involving personal connections (having either a positive or negative influence). If you ever suspect that you may have a COI, you should always tell the Editor.

Financial COIs can come about in a number of ways. For instance, if you receive funding from an organization that has a financial interest in your results, make sure that this is clearly stated in the cover letter and the acknowledgments section. If you personally have a financial interest in the outcome of the research, tell the Editor, even if the connection is not immediately apparent. Personal connection issues can range from, for instance, being friends with (or married to) one of the Associate Editors to having a long-standing feud with someone who might be chosen as a reviewer. In short, if your own or anyone else's professional judgment might be affected by a personal connection of any kind tell the Editor.

AUTHOR PENALTIES FOR UNETHICAL CONDUCT

Journal policies differ regarding the kinds of penalties they will apply when an author has been found to have acted unethically, whether it is a case of plagiarism, not reporting a COI, or manipulation of figures. Depending on the seriousness of the issue, the Editor may reject or retract the paper, or may inform the author's institution, leading to a formal inquiry, which can lead to dismissal in very serious cases. The author may be forbidden from submitting papers to the journal, either for a specific period or for a lifetime. Finally, in the US, the Editor may turn the investigation over to the Office of Research Integrity (ORI).

If the issue involves a dispute over authorship, this is also often turned over to the ORI for investigation and resolution; this may result in a delay to the publication of the paper in question and may even result in the authors' research programs being stopped until the issue has been resolved.

SUPPORT AND ADVICE ON ETHICAL ISSUES FOR EDITORS

When a case of unethical behavior, either by an author or a peer reviewer, comes to light, the Editor is often faced with difficult decisions. Clearly, any actions taken must be reasonably swift and also scrupulously fair. Many publishers have an ethics policy in place, with specific guidelines that Editors must follow; however, Editors can discuss general ethical issues with other Editors and also seek guidance and advice on specific cases if they are members of the Committee on Publication Ethics (COPE). This UK-based not-for-profit organization was launched in the late 1990s by a small number of medical Editors. Since then, membership has increased to over 6,000 individuals from all around the world, representing a wide range of scientific and medical specialties. They also represent all sectors of scientific publishing, from small association publishers to big international for-profits. COPE members are expected to adhere to a code of conduct for journal Editors. Therefore, when an author complains about the actions of an Editor, COPE may be asked to investigate if the Editor is one of their members (see next section).

UNETHICAL BEHAVIOR BY EDITORS

The most common issues that arise regarding unethical behavior among Editors involve COIs of one kind or another. The bottom line is that, provided the paper is a good match for the journal (and that there are no problems with backlogs or an excess of papers on a particular topic), Editors must judge each paper they receive based on the merits of the science and its presentation, and not on personal considerations regarding either the research, the authors, or the institution they work for. If an Editor has a personal connection with an author, that Editor must declare a conflict of interest and abstain from making any decisions on the paper. Similarly, if the Editor has a financial interest in the outcome of the research, the same rule applies. In addition, Editors must not allow personal biases connected with race, sex, religion, or nationality to influence their judgment of a paper in any way. Unfortunately, such biases are often subconscious and subtle and therefore hard to detect. Only by looking at an Editor's record over time is it possible to see such influences at work (see chapter 9, "Who Does What in Peer Review").

Two other ethical issues can arise in connection with Editors. First, science journals are most likely to have Editors who are experts in the scientific specialty covered by the journal, and who are often practicing scientists themselves. These Editors can therefore sometimes find themselves in the position of looking at a paper that relates to their own work in some way. The results may contradict the Editor's own findings or the paper may be in competition with the Editor's own paper on the same topic. However, Editors are in a position of trust; they must not use the information in submitted papers until after it has been published and must not delay its publication in any way in order to allow their own work to appear first.

The second ethical issue that may arise is where an Editor fails to maintain the anonymity of the peer reviewers or Associate Editor (in a few journals these remain anonymous as well). Reviewers need to be able to write their reviews frankly and honestly, and anonymity is particularly important where the review is critical of the paper. If the reviewer's name is revealed, this could do great damage to personal relations between the author and reviewer. Unfortunately, there have been occasions where an Editor, or a member of the editorial staff, has had a lapse in concentration and accidentally left a reviewer's name on a document that the author will see. Luckily, such occasions are rare. However, it would be highly unethical for an Editor to deliberately allow an author to discover the name of an anonymous reviewer without permission from the reviewer. On the other hand, it should be noted that reviewers do sometimes request that their name be made known to the authors of a paper, so that those authors can contact them with questions if they wish to. In such cases, it is perfectly appropriate for the Editor to reveal the reviewer's name.

The penalties for unethical behavior by Editors or reviewers vary greatly, depending on the seriousness of the issue. They may be reported to the institution or company that employs them, leading to an investigation and possible sanctions or dismissal. Alternatively, the investigation may be carried out by the ORI or an equivalent national body.

Any discussion of ethics is hampered by the fact that each individual's views on the topic are affected by their personal values and biases—what may seem like a clear-cut case of right or wrong to one person will be a question of "maybe or maybe not" according to the beliefs and experiences of another. For this reason, it is important that the institutions where scientists work should have clear ethics guidelines. Editors should also have a clear code of ethics to work with, so that authors and their papers are judged fairly, according to the guiding principles laid down by the publisher.

THE BOTTOM LINE

Like all other professionals, scientists are under pressure to succeed in their careers. However, scientists are under particularly heavy pressure, as many are tackling some of the most serious problems we face today, from infectious and genetically based diseases to global climate change, biodiversity losses, and rising pollution levels. The temptation to cut corners, either in the conduct of the research or its eventual publication, can be very strong but must be resisted at all costs. Researchers need to understand all the aspects of unethical behavior in science and science publishing, some of which can be quite subtle, so they can avoid these pitfalls and maintain the highest possible ethical standards in their work.

FURTHER READING

The Ethics in Science website (http://www.files.chem.vt.edu/chem-ed/ethics/index.html) provides links to many useful materials. The Council of Science Editors' *White Paper on Promoting Integrity in Scientific Journal Publications, 2009 Update* can be found at http://www.councilscienceeditors.org/files/public/entire_whitepaper.pdf.

CHAPTER TWELVE

Trends in scientific publishing

The future is as much about people and how they may be linked through content and activity as it is about the content itself. ROBERT HARINGTON, American Institute of Physics

Given the astonishing speed of advances in communication technologies, it seems almost silly to even attempt to talk about the "future" of publishing. Considering that the Internet is only thirty years old, can we realistically envision what will be available thirty years from now? Probably the best we can do is to look at what is happening today and guess at which of today's trends in scientific, technical, and medical (STM) publishing might persist over time and where they might lead. We end this book about what Editors want by looking at a handful of trends that we think will likely continue to affect STM publishing over the next decade and that we think are important for readers to keep on their radar screens. This list is by no means exhaustive and no doubt new technologies will emerge that will further change the publishing landscape, but for now we can only see what is on the current horizon.

THE BRUSSELS DECLARATION

A good place to start a discussion about current trends is with a brief overview of the Brussels Declaration, a document adopted in 2007 by a large group of publishers to make a stand in response to a heated debate that was going on at the time about the value of Editors and publishers and about the economic models that support the publication and dissemination of science (see appendix 5 for the text of the Brussels Declaration). Authors were (and still are) also thinking about the possibility of publishing systems where peer review would be an open, transparent, and ongoing process, and where there would be no costs to publishing or to accessing scientific literature.

In the Brussels Declaration, publishers point out that although the idea of completely free and open publishing is laudable, it is also unrealistic, unachievable, and, in fact, unwise. Publishers and Editors play roles that scientists do not. They organize and manage publications, monitor copyright and intellectual property issues, and launch new publications that reflect changes in the interests of researchers as well as changes in reading technologies. They also create, maintain, and build connections among data sets, publication archives, and other content. They also publish in multimedia and enable publication in both broad and narrow disciplines.

Editors and publishers do things that scientists cannot. If scientists spent the time developing the professional expertise necessary to do what Editors and publishers do, they would not be scientists: they'd be Editors and publishers. The Brussels Declaration lays out principles that appropriately set the stage for a discussion of future developments in how science is published, disseminated, and archived. Along with Editors and publishers, authors will all be challenged to keep up with these trends in the coming decades because, one way or another, they will impact how science is measured, published, and valued.

The concerns of publishers are interwoven with those of Editors on many fronts, but here we focus on some of the most prominent, namely changing financial models for publishing, new systems of peer review, alternative ways to measure impact, new technologies, the move towards data and content sharing, and public understanding of science.

COSTS WILL CHANGE BUT WILL REMAIN

In several places throughout this book, we've talked about the costs of publication. Most of the principles outlined in the Brussels Declaration relate to these costs: sustainable financial models are at the root of all the discussions about intellectual property, egalitarian access to information, and expert review of science. Unquestionably, the costs of publishing are changing rapidly and dramatically as new channels and innovative technologies for communicating science evolve. At this point, no one can predict what the next hot new tool will be and how much it will cost to develop and maintain. However, as new technologies are created and with them new business models for covering costs, authors are certain to be called on to pay at least some of those expenses. It has taken more than a decade of discussion about the pros and cons of open access (OA) publishing for publishers, Editors, and readers to acknowledge that perhaps the most important aspect of the proliferation of OA journals is that they reflect the success of economic models that shift the burden of cost from libraries and readers to authors. Authors, and the organizations that fund them, will very likely be called upon to cover other costs as well, including costs we may not yet be able to predict.

The costs of publishing, distributing, linking, and archiving science are in flux, and they are likely to remain so for the foreseeable future. Today, publishers, authors, librarians, governments, not-for-profit organizations, philanthropic institutions, and others are participating in experiments with the economics of publishing. In the coming decades, you will likely be paying fees where before there were none while at the same time seeing other costs, such as page charges or fees for color figures, disappear. You may see new fees levied by libraries, required in grant applications, mandated for archiving, or collected to cover linking technologies. Unlike the economic models of the past, new models to support scientific publishing will likely become a moving target, changing as quickly as the supporting technologies. As a member of the scientific community, you will need to read the fine print of agreements when you apply for or receive grants, when you publish, when you access information online or elsewhere, to be sure you understand the required payments and what the costs will cover.

PEER REVIEW WILL CHANGE BUT WILL REMAIN

As mentioned in chapter 9, peer review is widely acknowledged as an imperfect system; nonetheless, some form of expert review will always remain as a gateway to publication. The research being debated at the International Congress on Peer Review and Biomedical Publication, held every four years, is a clear acknowledgment that there are faults in the system, but efforts are under way to understand and manage these issues. To find evidence of these flaws, just keep an eye on your favorite news source to see regular reports on research that has been retracted because of problems that should have been caught in peer review. Papers are retracted for a worrying array of reasons, including problems in the statistical analyses or experimental design, or due to plagiarism or falsified data.

Who carries out the peer review, whether they remain anonymous or not, how transparent the process is, how open the content of the reviews becomes, and when it happens are all elements open to change. Authors will no doubt continue to write about the flaws of the review system and to offer up alternatives. In turn, every article advocating a new kind of review systems will likely be followed by a cascade of rebuttals, highlighting the flaws in those alternatives. In the meantime, everyone involved in publishing science will still be working to improve the review process by creating innovative online mechanisms for evaluating and commenting on new research.

Being part of these multiple channels of conversation may well become a requirement for researchers who want to establish a reputation in their field. At the very least, you will need to be aware of the most popular networks for discussion and information sharing. But for the time being, peer review that is orchestrated by Editors, carried out by subject matter experts, and supported by technology that costs money is the best way we have of evaluating science submitted for possible publication.

MEASURES OF IMPACT WILL CHANGE BUT WILL REMAIN

As discussed in chapter 6, some Editors, publishers, and authors acknowledge the dangers in the misuse of impact factors and agree with its inventor's assessment that this metric has become a mixed blessing. We need more robust ways to assess the impact and importance of researchers, the articles they write, and the journals in which they publish. New measures are regularly proposed, so that today, Editors, publishers, biometric researchers, and authors have a smorgasbord of alternative metrics to choose from: the eigenfactor, the *h*-index, the G-index, recipes for calculating citation half-lives, aggregated citations, and formulas for page ranking created by Google, Scopus, and similar organizations. New ways of measuring usage and influence have been exploding onto the scene since publications went online, and it is still too early to tell which of these measures will take root and be used by researchers to assess the latest papers or by academic promotion committees to evaluate the publication lists of researchers.

New metrics will continue to be developed and maybe one day a metric will be created that will take the place of the journal impact factor we use now. Today, however, there is no single new superpower metric on the horizon, but a new one may well appear, as new technologies and web-based social networks evolve.

TECHNOLOGY WILL CHANGE, AND CHANGE, AND CHANGE

Technologies that impact the working lives of researchers are evolving at an extraordinary rate; and again, what is a trendy tool today may be obsolete tomorrow. For example, the pace of development of application software (apps) programs is currently moving ahead at breakneck speed and an enormous number of apps are being designed specifically to support researchers and authors in the sciences (see sidebar 12.1). Apps are popular because they are usually affordable, are designed to meet specific needs, and are highly portable. Some publishers are creating apps that allow users to browse journal content, read abstracts, and view or download full text. Others are putting reference materials into application formats, allowing users to quickly access comprehensive information, including useful resources for clinical diagnosis and practice. Applications for generating output that will end up in published papers are also becoming available, ranging from apps for decoding genetic materials to interactive tools for mapping clusters of stars. The assortment and power of apps will undoubtedly continue to grow as the underlying technologies evolve, the software tools for creating new applications improve, and app developers better understand what researchers are willing to pay for.

SIDEBAR 12.1

The evolving role of mobile apps in STM publishing

SINAE PITTS, PH.D.
CEO, Amphetamobile

The landscape of mobile applications is now rapidly expanding with new tools that allow readers to discover, consume, and produce new content. This evolution in technology is leading to truly new science as well as to groundbreaking new channels of communication.

CONTENT DISCOVERY

To keep up with the increased flow of information, many journals now have apps that users can configure to notify them when new articles or issues can be accessed and even when articles containing specific keywords have come online. Through mobile device connectivity, readers can more easily and quickly keep up with the latest developments in their fields and engage in this critical task during times that used to be only minimally productive, such as when they are commuting or while waiting for a short lab experiment to finish. The increasing use of social networks and social bookmarking is also leading to new efficiencies, as users share content with like-minded peers and take advantage of the immediacy and community aspects of being linked through mobile access. All of these innovations have raised new questions for prospective authors, including how the impact of their work can be measured. Authors are increasingly aware of how quickly their papers can go from acceptance to the hands of readers, with delivery to mobile devices now the fastest route.

CONTENT CONSUMPTION AND AVAILABILITY

It took only a few years for readers to switch from reading content in print to primarily reading online, and mobile reading is projected to eclipse desktop browser reading even more rapidly. Mobile apps return to readers the convenience of consuming journal content where and when they want, and mobile devices are being adapted to enable doctors and researchers to invent and use remarkable new tools in amazing new ways. Scientists can now use their personal and institutional mobile devices to look up, search for, read, absorb, and use information far more effectively and efficiently than they could when they were tethered to a desk. In addition, mobile devices such as smartphones are now available in all corners of the world, and mobile connectivity is readily available and less expensive than "traditional" access through desktop computers and Ethernet services, particularly in developing countries.

CONTENT PRODUCTION

For online and mobile consumption, authors now have the opportunity to create new kinds of content and are doing so in growing numbers as they abandon the limitations of print. Users continue to demonstrate an appetite for full-length, long-form content, and apps and devices are getting better and better as new tablet technology comes to market. Within the foreseeable future, journals for online- or mobile-only reading will allow authors and readers alike to take advantage of the powerful communication potential of high-resolution color, audio, and video content.

Mobile devices are also becoming more effective in supporting sophisticated content-generating apps, and new tools allow readers to more deeply interact with content, to collaborate, comment, and manage references and copyrighted materials. The next generation of apps will allow even more robust content creation and uses that will both mimic traditional forms of desktop publishing and also empower novel forms, such as freehand drawing and annotations with digital ink, photo and video capture, and audio recordings.

NEW TRADITIONS IN STM COMMUNICATION

Scientific publications have traditionally allowed researchers to promote their own work as well as to further their own knowledge and have also allowed people to connect with peers, build relationships, learn, and teach. Mobile devices and applications have strengthened, and will continue to support, these critical activities and will likely also foster deeper interactions and stronger ties with distant users. Mobile apps will do a lot more than just facilitate traditional peer review. They will enhance scientific discovery by allowing a continuous stream of written and multimedia content to push the traditional idea of "the article" into new shapes and with them our understanding of what constitutes reading and authorship.

The breakneck speed with which apps are being created and sold is evidence of the voracious appetite that readers and writers have for the immediacy, convenience, and creative potential that mobile apps bring to science communication. New ideas and new knowledge will be effective if they can be widely disseminated, and today, mobile apps are proving to be mightier than the pen.

Electronic readers (or eReaders) are another rapidly expanding new technology. Dozens of eReaders are now on the market, and by the time you have purchased one and gotten accustomed to using it, a better, faster, and cheaper one will be available. Although many eReaders use similar electronic file formats, these standards are far from being final. ePub, developed by the International Digital Publishing Forum, seems to be a strong contender for becoming the accepted standard, free, and open format for digitizing materials for eReaders. However, publishers have not yet solved all the associated challenges, such as putting mathematical equations, complex images, and linking technologies into eReader formats.

Many researchers, particularly younger ones, are interested in eReader technologies because they want to read their favorite journals and want others to read their papers on these devices. However, relatively few journals and books are available on eReaders in any form other than PDFs, in part because publishers have not yet found ways to get a return on investment for putting scholarly information into ePub formats. Large publishers may be able to get their content into electronic formats that can be read on most eReaders, but smaller publishers, particularly not-for-profit society publishers, may not have the money, the staff, or the expertise to do so, especially since the technologies are changing so quickly. The movement of text from online to eReaders is yet another part of the evolution of publishing that will move ahead slowly, driven by cost.

CONTENT, CONTENT EVERYWHERE

Another trend already impacting many authors involves data and content sharing. Here, we use the term "content" to refer to any data, video, audio, or other materials that might be associated with a scientific paper and the term "data" to refer specifically to data sets. Many organizations that fund research are introducing new regulations concerning content sharing and in line with these regulations, a growing number of journals are requiring authors to submit the data on which a paper is based. At the same time, however, a few journals are refusing to accept supplementary content because of the cost and complexity of maintaining data archives.

Requirements for content sharing vary enormously from field to field and from publisher to publisher, differing in terms of whether or not supplemental content must be shared at all, and if so, who should be allowed to access it, where the material should be archived, and who should pay the associated costs. In some situations, authors may be strongly disinclined to share data, as these may be the basis of a new product, drug, or patent, and sharing that information would cause them to lose their competitive edge or give away the basis for trade secrets. In other situations, requirements for data sharing may be unclear, or may have implications for copyright or ownership of intellectual property. However, whether required or not, most Editors and authors agree that access to supplemental content does allow reviewers and readers to more thoroughly evaluate research and that requiring access to content is beginning to change the sense of "ownership" of data that researchers have had for centuries.

The movement to provide access to content through open access journals and data repositories underscores the globalization of scientific research and the increase in international collaboration in science. A number of organizations and resources that provide the necessary infrastructure for data archiving and linking are being launched to foster collaboration, not only within disciplines (e.g., the Geological Society of America's Data Repository at www.geosociety.org/pubs/drpint.htm#aboutdrp), but across disciplines as well (e.g., Alliance for Permanent Access, www.alliancepermanentaccess.org or Parse.Insight, http://www.parse-insight.eu). Further efforts to extend the flexibility of other media, including video-journals (such as the *Journal of Visualized Experiments*), mobile and tablet applications, and other tools are also being developed (see sidebar 12.2).

SIDEBAR 12.2

An exemplar of new media in STM publishing

MOSHE PRITSKER, PH.D.
CEO, Editor-in-Chief, and Cofounder, *Journal of Visualized Experiments*

Biomedical research has reached a level of complexity that is matched only by the complexity of the living species under investigation. Yet despite the complexity and rapid advancement of scientific research itself, scientific communication still relies heavily on traditional text journals. The format of these journals has remained practically unchanged for the past 200 years. The inherent limitations of the text format include:

- *the requirement for authors to represent complex experimental studies in writing*
- *the requirement for readers to correctly interpret complex descriptions*
- *variation in employed terminologies.*

As a result of these limitations, knowledge transfer is often inefficient, as reflected in the fact that it is especially difficult for other scientists to reproduce newly published experimental studies. Thus, a lot of time in biomedical research is consumed in repetitive attempts to establish and use laboratory techniques and procedures described in the scientific literature. This has become a never-ending process for biological scientists as technologies in this fast-growing field undergo significant changes every few years (e.g., genomics and proteomics). Economically, the time and resource-consuming process of training and retraining to carry out techniques and procedures represents a critical "bottleneck" for biomedical research and drug discovery.

To increase the efficiency of knowledge transfer, Journal of Visualized Experiments (JoVE, *www.jove.com) was founded in late 2006 by Klaus Korak and me, both postdoctoral scientists at Harvard Medical School, and Nikita Bernstein, a computer programmer.* JoVE *represents a novel approach to scientific publishing, using video format to publish step-by-step demonstrations of advanced experimental studies performed in the laboratories of leading research institutions such as Harvard, MIT, Yale, Stanford, Berkeley, and many others. Visualization through video greatly facilitates the understanding and efficient reproduction of experimental techniques, which in turn greatly increases transparency and efficiency in biomedical research. To quote a commonly used proverb, "A picture is worth a thousand words."*

JoVE *became the first video-journal accepted for indexing in MEDLINE and PubMed, the official repository of scientific journals maintained by the National Library of Medicine. To effectively film at universities and biotech companies around the world,* JoVE *has established a geographically distributed network of video professionals covering big cities in many countries, including the US, Canada, UK, Germany, Sweden, Israel, Australia, and Japan. As of 2011,* JoVE *has published over 1,200 video articles, is publishing 50 new articles each month, and is serving as an example of how innovative use of technologies can foster the evolution of scientific information transfer. As technologies develop and Internet access expands globally in the coming decades, new and powerful ways to leverage new media will surely evolve, to communicate scientific methods, studies, results, and paradigms.*

So, although authors will undoubtedly encounter requirements for data deposition, content sharing, and linking initiatives, where these developments lead will differ among specialties, funders, research communities, and publishers (see sidebar 12.3). As an author, you will need to keep abreast of data-sharing requirements, opportunities, and costs in your area of research and make sure that you conduct your research and archive your data in line with these requirements.

PUBLIC INTEREST IN SCIENCE AND POLICY

As a source of new knowledge that can shape public opinion and behavior, scientists bear an increasing responsibility to be able to communicate clearly, concisely, and accurately to the public. Many scientists will admit that communicating with nonscientists is not their strongest skill and that they are most comfortable discussing their work within their own research communities, through publication and conference presentations. Only a few scientists enjoy the task of translating their findings for public consumption, while others dislike having to simplify complex ideas for nonscientific audiences or believe that the significance of their findings won't register with those who lack a scientific background.

Nonetheless, scientists will continue to be called upon to clearly and concisely articulate the implications of their own findings, as well as the work of others, although how decision makers and the public respond to the opinions of experts varies greatly. In some cases, people take scientific findings immediately to heart and change their behaviors, as when they are told about studies showing a certain medicine or medical procedure may be unsafe. However, in other areas, decision makers and the public are often unconvinced by what scientists tell them, regardless of how compelling the evidence is, as with studies related to climate change. Given that many new findings have profound global implications, we believe that all scientists should develop the skills necessary to effectively translate their findings into language that nonspecialists can understand. When scientists use clear language to articulate their perspectives on science related to human and environmental health, they have the potential to sway public opinion and influence the outcome of policy and legal debates. Having a strong and clear voice in the public realm can also help scientists to be more effective in fostering the accountability of researchers, the transparency of funders, and the activities of corporations in areas of social responsibility. Many scientific organizations and publishers provide training and resources to help scientists develop strong communications skills, reflecting their efforts to guide the general public towards more sustainable and healthy behavior.

SIDEBAR 12.3

The future of publishing

ROBERT M. HARINGTON, D.PHIL.
Publisher, American Institute of Physics

As we look ahead to the future of publishing, we should not be too concerned with what appears to be the crumbling foundations of scholarly publishing on which we now stand. Yes, print is on its way out. Yes, there is a tectonic shift in the balance of power among content stakeholders, authors, Editors, publishers, and readers. Yes, the fundamental business models of publishing are shifting. The world's academic institutions (certainly in the publishers' traditional markets) and their libraries are in economic turmoil. In the end, however, we must focus on the fact that content *has intrinsic value and the* relationships *in and around content provide a social framework, not only for creating new science but also for powering the business of publishing. Knowing that content and relationships will remain at the core of scholarly communication, we can start to look at the future of publishing by asking what a publisher does to add value. I hope that readers will be as excited as I am by the wealth of opportunities that scientific publishing brings us. As the caretakers of Editors, publishers provide the housing and tools for editorial work, and what publishers do is intrinsically tied to what Editors do and what they want and need.*

The publisher has a number of roles that enable scholars to teach and research more effectively. Anecdotal evidence suggests that the currency of scholarship—the academic paper—is not going away anytime soon. Publishers are concerned with quality of content and of publishing operations, focusing on efficiency, simplicity, and transparency. The publisher's job is to ensure the right blend of speed to publication, quality editing, and use of tools that allow readers to easily consume content, from finding what they want to the deep reading of content with links to the broadest possible range of information related to the paper.

First, publishers must enable researchers and educators to find what they need when they need it. We must ensure not only that research is semantically enabled but that serendipity continues to play a part in connecting readers to intriguing and novel material. Publishers must also pursue the development of tools that will allow scholars to meet, share content, and cultivate ideas in a wider online environment. We need to help authors create and collaborate in real time and enable them to monitor the publishing process at every stage. Through linking technologies, publishers help readers connect to authors, content, and each other, and in doing so provide a true collaborative view of the scholarly endeavor.

The key to evolving online presentation of content is to move away from the idea of a fixed and comprehensive publisher platform, and instead adopt an understanding of publishing as a range of adaptable tools and processes that can easily change to suit the changing needs of authors and readers.

How content is used and how widely it can be accessed will be one key measure of a publisher's success. They will therefore need to become more involved in the development and use of mobile devices, which will play an increasing role in accessing data and interacting with collaborators and students. Although libraries will remain the guardian and reflection of an institution's academic quality, the new sanctuary of scholarship will be portable, global devices that allow academics to access information, collaborate, read, and write. As publishers consider the future, we are also addressing the new complexities of copyright, addressing it not only as a protective veil but as an opportunity to encourage collaboration and innovation—which is really what copyright law was always supposed to be. Increasingly, we will need to let go of copyright and let it sit with the author or in the public domain.

Publishers are already seeing tremendous opportunities in the disassembling and repurposing of content and tools that are already being developed to connect content, people, and the business drivers backing them up. Developing effective business models is key to the future of publishing, and developing online environ-

ments that foster scholarly collaboration will also draw together people and the products they want and need. We know that the current financial models used to generate revenue, such as charges for access to content, are not yet dead. While the traditional print journal subscriptions will probably disappear over time, publishers will increasingly be able to power the financial engine of publishing through the packaging of content for sale, at individual levels through mobile devices, or at higher levels, through subscriptions to institutions, or even by licensing content at a national level. In addition, the psychological need of like-minded scholars to form societies will continue to be a relevant business model for publishers who, through innovative collaborative tools, will bring new concepts of membership to the scholarly community.

Publishers will continue to play essential roles in the lives of researchers and educators as long as they continue to add value to content. Yes, the successful academic publisher of the future will need to have high-quality publishing operations, but they will also need to provide authors and readers with increasingly refined tools that will help researchers to be more productive in their daily routines. If publishers can provide authors and readers with intuitive ways to search for information and powerful tools to reach deeper into content, meaning, connections, and relationships with peers, then there is a bright future ahead for all.

THE BOTTOM LINE

We began this book by restating the maxim that all academics and researchers know by heart—publish or perish—and we said that this remains true and will be for the foreseeable future. However, the technologies of text are changing so dramatically that many of today's popular tools for publishing and disseminating science were unimaginable even ten years ago. What new technologies will be developed and take root in the next ten years is anyone's guess. Scientists will still need to publish their work and have it vetted by a community of experts, but how that will be done remains just over the edge of the horizon.

APPENDIX 1

Resources for improving science writing

Online resources for improving science writing

Academic resources

Capital Community College Foundation Guide to Grammar and Writing	grammar.ccc.commnet.edu/grammar/
Colorado State Open Access Writing Studio	writing.colostate.edu/
Online English Grammar Resources	www.edufind.com/english/grammar/grammar_topics.php
Paradigm Online Writing Assistant	www.powa.org/
Purdue University Online Writing Lab	owl.english.purdue.edu/
University of North Carolina at Chapel Hill: The Writing Center	writingcenterunc.edu
University of Ottawa: Writing Centre (Note: this site uses Canadian spelling)	www.arts.uottawa.ca/writcent/hypergrammar/grammar.html
Penn State University course site for Writing Guidelines for Engineering and Science Students	www.writing.engr.psu.edu/

US government resources

HHS Centers for Medicare & Medicaid Services: Toolkit for Making Written Material Clear and Effective	www.cms.gov/WrittenMaterialsToolkit/
NIH: Communicating Research Intent and Value in NIH Applications	grants.nih.gov/grants/plain_language.htm

NIH Library: Writing Center	nihlibrary.campusguides.com /WritingCenter
NIH: Clear Communication	www.nih.gov/clearcommunication/plainlanguage.htm
Plain Language.gov	www.plainlanguage.gov/

Other resources

Program for Readability in Science and Medicine (PRISM)	www.grouphealthresearch.org /capabilities/readability/readability_home.html
Clinical Chemistry Guide to Scientific Writing	www.aacc.org/publications/clin_chem /ccgsw/pages/default.aspx
Current Medical Research & Opinion	www.cmrojournal.com/ipi/ih /MPIP-author-toolkit.jsp
Online English Grammar Resources	www.edufind.com/english/grammar /grammar_topics.php

Manuals and style guides

AMA Style Manual Committee. 2007. *AMA Manual of Style: Official Style Manual of the American Medical Association* (10th ed.). NY: Oxford University Press.

Bates R *et al.* (Eds.). 1995. *Geowriting: A Guide to Writing, Editing, and Printing in Earth Science* (5th ed.). Alexandria, VA: American Geological Institute.

Chicago Style Manual Committee. 2010. *The Chicago Manual of Style* (16th ed.). Chicago: University of Chicago Press.

Coghill A and Garson L. 2006. *The ACS Style Guide: Effective Communication of Scientific Information*. Washington, DC: American Chemical Society.

Council of Science Editors. 2006. *Scientific Style and Format: The CSE Manual for Authors, Editors, and Publishers* (7th ed.). NY: Rockefeller University Press.

Patrias K. 2007. *Citing Medicine: The NLM Style Guide for Authors, Editors, and Publishers* (2nd ed.). Wending DL, Technical Editor. Bethesda, MD: National Library of Medicine. Available at www.nlm.nih.gov/citingmedicine.

Rabinowitz H and Vogel S. 2008. *The Manual of Scientific Style: A Guide for Authors, Editors, and Researchers*. NY: Academic Press.

Turabian K *et al.* 2007. *Manual for Writers of Research Papers, Theses, and Dissertations: Chicago Style for Students and Researchers* (7th ed.). Chicago: University of Chicago Press.

U.S. Government Printing Office Style Manual: An Official Guide to the Form and Style of Federal Government Printing (30th ed.). Washington, DC: US Government Printing Office.

Weiss E H. 2005. *The Elements of International English Style: A Guide to Writing Cor-*

respondence, Reports, Technical Documents, and Internet Pages for a Global Audience. Armonk, NY: ME Sharpe.

Textbooks on writing, editing, and publishing

Belcher W. 2009. *Writing Your Journal Article in Twelve Weeks: A Guide to Academic Publishing Success.* NY: Sage.

Briscoe M. 1996. *Preparing Scientific Illustrations: A Guide to Better Posters, Presentations, and Publications* (2nd ed.). NY: Springer.

Cargill M and O'Connor P. 2009. *Writing Scientific Research Articles: Strategy and Steps.* Hoboken, NJ: Wiley.

Day R and Gastel B. 2006. *How to Write and Publish a Scientific Paper* (6th ed.). Phoenix: Oryx.

Day R and Sakduisku N. 2011. *Scientific English: A Guide for Scientists and Other Professionals.* Phoenix: Oryx.

Divan A. 2009.*Communication Skills for the Biosciences: A Graduate Guide.* Oxford: Oxford University Press.

Fishman, S. 2008. *The Copyright Handbook: What Every Writer Needs to Know.* Berkeley: Nolo.

Fogarty M. 2009. *The Grammar Devotional: Daily Tips for Successful Writing from Grammar Girl.* NY: MacMillan.

Huckin T and Olsen L. 1991. *Technical Writing and Professional Communication: For Nonnative Speakers of English* (2nd ed.). NY: McGraw Hill. (See also Olsen and Huckin, 1991.)

Humphrey J *et al.* 2009. *Style and Ethics of Communication in Science and Engineering.* San Rafael, CA: Morgan and Claypool.

Isaacs A and Daintith J. 2009. *New Oxford Dictionary for Scientific Writers and Editors.* Oxford: Oxford Univesity Press.

Jones D. 1998. *Technical Writing Style.* Boston: Allyn & Bacon.

Knapp M and Daly J. 2004. *A Guide to Publishing in Scholarly Communication Journals.* Mahwah, NJ: Earlbaum.

Korner A. 2008. *Guide to Publishing a Scientific Paper.* NY: Routledge.

Lang T. 2010. *How to Write, Publish, and Present in the Health Sciences: A Guide for Clinicians and Laboratory Researchers.* Philadelphia: American College of Physicians Press.

Lipton R. 2007. *A Practical Guide to Information Design.* NY: Wiley.

Olsen L and Huckin T. 1991. *Technical Writing and Professional Communication.* (2nd ed.). NY: McGraw Hill.

Murphy A. 2010. *New Perspectives on Technical Editing.* NY: Baywood.

Penrose A and Katz S. 1998. *Writing in the Sciences: Exploring Conventions of Scientific Discourse.* NY: St. Martins.

Ribes R *et al.* 2009. *English for Biomedical Scientists.* London: Springer.

Schriver K. 1997. *Dynamics in Document Design: Creating Texts for Readers.* NY: Wiley.

Schultz D. 2009. *Eloquent Science: A Practical Guide to Becoming a Better Writer, Speaker, and Atmospheric Scientist*. Boston: American Meteorological Society.

Stilman A. 2010. *Grammatically Correct: The Essential Guide to Spelling, Style, Usage, Grammar, and Punctuation*. Cincinatti: F+W Media.

Stim, R., 2010. *Getting Permission: How to License & Clear Copyrighted Materials Online & Off*. Berkeley: Nolo.

Stuart J and Scott J. 2009. *Study and Communication Skills for the Biosciences*. Oxford: Oxford University Press.

Williams J. 1995. *Style: Toward Clarity and Grace.* Chicago: University of Chicago Press.

Zeiger M. 2000. *Essentials of Writing Biomedical Research Papers* (2nd ed.). NY: MacGraw-Hill.

APPENDIX 2

Databases with free access to articles or abstracts

Name of database	*Subject area*	*URL*
Agricultural Sciences and Technology Database (AGRIS)	agriculture	www.ntis.gov/products/agris.aspx
Analytical Sciences Digital Library (ASDL)	analytical sciences	www.asdlib.org/
arXiv	Physics, mathematics, computer science, biology, statistics	arxiv.org/
Astrophysics Data System	astrophysics, geophysics, physics	adswww.harvard.edu/
$Chem_xSeer$	chemistry	chemxseer.ist.psu.edu/
Circumpolar Health Bibliographic Database (CHBD)	medicine	www.aina.ucalgary.ca/chbd/
$CiteSeer^X$	computer science	citeseerx.ist.psu.edu/about/site
Directory of Open Access Journals (DOAJ)	journals	www.doaj.org/
Google Scholar	multidisciplinary	scholar.google.com/
JournalSeek	multidisciplinary	journalseek.net/
OAIster	multidisciplinary	www.oclc.org/oaister

Name of database	*Subject area*	*URL*
PubMed	medicine	www.ncbi.nlm.nih.gov/pubmed
PubChem	chemistry	pubchem.ncbi.nlm.nih.gov/
Science.gov	multidisciplinary	www.science.gov/
WorldCat	multidisciplinary	www.worldcat.org/

APPENDIX 3

Presubmission checklist

Check all of the following items one final time before you submit your manuscript.

General

Check back issues of the journal and instructions to authors and make sure that:

- ☐ Your paper fits within the stated scope of the journal.
- ☐ You have followed the journal's formatting and style instructions.
- ☐ Your manuscript is within or near the stated word limit.
- ☐ You've included no more than the number of references allowed.
- ☐ You've formatted your in-text citations and reference list in the required format.
- ☐ You've done a final spelling and grammar check.
- ☐ You have enough money to pay for page charges and other costs.
- ☐ You have checked the information on where and how to submit.
- ☐ You've had your English checked by a native speaker of English (if you are a non-native speaker/writer).
- ☐ You've written in short, straightforward sentences and avoided jargon whenever possible or added definitions or explanations as appropriate.
- ☐ You've spelled Latin and other names, place names, and chemical and drug names correctly and consistently throughout the text.
- ☐ Your numbers are consistent throughout the text, figures, and tables, and you have explained any apparent discrepancies.
- ☐ You've spelled out all acronyms when they first appear.
- ☐ You've added line numbers to the manuscript unless directed not to do so in the instructions to authors.

Authors

Make sure that:

- ☐ All coauthors have seen and approved the final version of the paper.
- ☐ You have all the required forms from each author (e.g., disclosure forms, copyright forms).

- ☐ You've listed everyone appropriately in the author and acknowledgment sections.
- ☐ All the correct contact details for all authors are on the front page of the manuscript.
- ☐ All the correct information is included for the corresponding author.

Cover letter

Make sure that:

- ☐ The cover letter is no more than one to one and a half pages.
- ☐ You've used the exact name of the correct journal in the letter.
- ☐ You've included compelling information about your manuscript in the letter.
- ☐ You have included all necessary information about conflicts of interest regarding either yourself or your coauthors in the cover letter.

Citations

Make sure that:

- ☐ All the references you cite in the text are included, properly spelled, with correct dates, and properly formatted in the reference list.
- ☐ Every listing in your reference list appears somewhere in the text.
- ☐ All citation details are correct.
- ☐ All Internet links cited in the text or in the references are correct and functional and you've included a "viewed" date.
- ☐ You've included and checked digital object identifiers (DOIs) wherever relevant (test them at www.crossref.org).

Figures

Study instructions to authors to make sure that:

- ☐ You have included no more than the maximum number of figures.
- ☐ You followed instructions for use of color and can afford color charges if required.
- ☐ Your files are within maximum or minimum size limits.
- ☐ Your files are all in accepted file formats (e.g., .eps, .jpg, .gif, .tif).
- ☐ All figures are the latest versions.
- ☐ All photos and graphics are at the resolution the journal requires for submission.
- ☐ You can supply all figures at the resolution needed for publication.
- ☐ You have permissions from the copyright holder (usually the publisher) for all figures that did not originate with you.
- ☐ All photos and graphics are at the correct sizes, and in the journal's preferred fonts and font sizes.
- ☐ All figures mentioned in the text are ready for submission.
- ☐ All the figures you plan to submit are actually mentioned in the text.
- ☐ All figures are referenced in numerical order in the text.
- ☐ All figure legends are listed wherever the instructions to authors specify they should be.

Tables

Study instructions to authors and make sure that:

- ☐ All tables are the very latest version.
- ☐ You have the necessary permission from the copyright holder (usually the publisher) for all tables that did not originate with you.
- ☐ All tables mentioned in the text are supplied with the manuscript and vice versa.
- ☐ All tables are referenced in numerical order in the text.

APPENDIX 4

Free and low-cost image resources

Free US Government photo resources

www.usa.gov/Topics/Graphics.shtml

Bureau of Land Management (BLM): natural areas, cultural and archaeological sites
www.blm.gov/wo/st/en/bpd.html

National Oceanic and Atmospheric Administration (NOAA): marine and coastal images, weather-related images, and more
www.photolib.noaa.gov

United States Department of Agriculture, Agricultural Research Service (USDA ARS): animals, plants, fruits, vegetables, crops, education, field research, lab research
www.ars.usda.gov/is/graphics/photos

Centers for Disease Control and Prevention: human and animal diseases, environmental health, natural disasters, bioterrorism, electron microscope imagery
phil.cdc.gov/phil/home.asp

US Fish and Wildlife Service National Digital Library: wildlife, plants, habitats, recreation, people, fisheries, training and education outreach, historical photos
www.fws.gov/digitalmedia

United States Geological Survey (USGS) Multimedia Gallery: historic and modern topographic maps, aerial and satellite images, and other technologies
gallery.usgs.gov

United States Forest Service: wildlife, fish, wildflowers, and environmental education photographs
www.fs.fed.us/photovideo

Inexpensive online photolibraries

www.iStockphoto.com

www.bigstockphoto.com

www.123rf.com

www.morguefile.com (lots of free photos and links to other free and cheap photolibrary sites)

APPENDIX 5

The Brussels Declaration

In 2007, sixty major publishing organizations adopted the Brussels Declaration. A pdf, including a list of signatories, is available at www.stm-assoc.org/brussels-declaration.

Brussels Declaration on STM Publishing

by the international scientific, technical and medical (STM) publishing community as represented by the individual publishing houses and publishing trade associations, who have indicated their assent below

Many declarations have been made about the need for particular business models in the STM information community. STM publishers have largely remained silent on these matters as the majority are agnostic about business models: what works, works. However, despite very significant investment and a massive rise in access to scientific information, our community continues to be beset by propositions and manifestos on the practice of scholarly publishing. Unfortunately the measures proposed have largely not been investigated or tested in any evidence-based manner that would pass rigorous peer review. In the light of this, and based on over ten years experience in the economics of online publishing and our longstanding collaboration with researchers and librarians, we have decided to publish a declaration of principles which we believe to be self-evident.

1. **The mission of publishers is to maximise the dissemination of knowledge through economically self-sustaining business models.** We are committed to change and innovation that will make science more effective. We support academic freedom: authors should be free to choose where they publish in a healthy, undistorted free market
2. **Publishers organise, manage and financially support the peer review processes of STM journals.** The imprimatur that peer-reviewed journals give to accepted ar-

ticles (registration, certification, dissemination and editorial improvement) is irreplaceable and fundamental to scholarship

3. **Publishers launch, sustain, promote and develop journals for the benefit of the scholarly community**
4. **Current publisher licensing models are delivering massive rises in scholarly access to research outputs.** Publishers have invested heavily to meet the challenges of digitisation and the annual 3% volume growth of the international scholarly literature, yet less than 1% of total R&D is spent on journals
5. **Copyright protects the investment of both authors and publishers.** Respect for copyright encourages the flow of information and rewards creators and entrepreneurs
6. **Publishers support the creation of rights-protected archives that preserve scholarship in perpetuity**
7. **Raw research data should be made freely available to all researchers.** Publishers encourage the public posting of the raw data outputs of research. Sets or sub-sets of data that are submitted with a paper to a journal should wherever possible be made freely accessible to other scholars
8. **Publishing in all media has associated costs.** Electronic publishing has costs not found in print publishing. The costs to deliver both are higher than print or electronic only. Publishing costs are the same whether funded by supply-side or demand-side models. If readers or their agents (libraries) don't fund publishing, then someone else (e.g. funding bodies, government) must
9. **Open deposit of accepted manuscripts risks destabilising subscription revenues and undermining peer review.** Articles have economic value for a considerable time after publication which embargo periods must reflect. At 12 months, on average, electronic articles still have 40–50% of their lifetime downloads to come. Free availability of significant proportions of a journal's content may result in its cancellation and therefore destroy the peer review system upon which researchers and society depend
10. **"One size fits all" solutions will not work.** Download profiles of individual journals vary significantly across subject areas, and from journal to journal

References

Alberts B. 2010. Promoting scientific standards. *Science* **327** (5961): 12. DOI: 10.1126/science.1185983

Anonymous. 2006. Overview: Nature's peer review trial. *Nature*. DOI:10.1038/nature05535

Bloom FE. 2000. Unseemly competition. *Science* **287** (5453): 689. DOI: 10.1126/science.287.5453.589

Brereton R. 2010. Authorship order in scientific papers. Suite101.com: http://www.suite101.com/content/authorship-orders—in-scientific-papers-a256157

Cho MK *et al.* 2006. Lessons of the stem cell scandal. *Science* **311** (5761): 614. DOI: 10.1126/science.1124948

Davidoff F. 2004 Improving peer review: who's responsible? *British Medical Journal* 328: 657. DOI: 10.1136/bmj.328.7441.657

Dellavalle RP *et al.* 2007. Frequently Asked Questions Regarding Self-Plagiarism: How to Avoid Recycling Fraud. *Journal of the American Academy of Dermatology* 57: 527. DOI: 10.1016/j.jaad.2007.05.018

Garfield E. 2000. Use of *Journal Citation Reports* and *Journal Performance Indicators* in measuring short and long term journal impact. *Croatian Medical Journal* **41**: 368–374.

Garfield E. 2006. The history and meaning of the journal impact factor (reprinted). *Journal of the American Medical Association* **295** (1): 90–93.

Gargouri Y *et al.* 2010. Self-selected or mandated, open access increases citation impact for higher quality research. *PLoS ONE* **5**: e13636. DOI:10.1371/journal.pone.0013636

Gunsalus, C. K. 1997. Ethics: sending out the message. *Science* **276** (5311): 335. DOI: 10.1126/science.276.5311.335

Harzing AW. 2011. Publish or Perish. Available at www.harzing.com/pop.htm.

Havens K. 2008. Bigwigs as coauthors: a response to Leimu *et al. Frontiers in Ecology and the Environment* **6**: 522–523.

Hirsch JE. 2005. An index to qualify an individual's scientific research output. *Proceedings of the National Academies of Science, USA* **102:** 16569–16572.

Interlandi J. 2006. An unwelcome discovery. *New York Times* (Sunday magazine). Oct 22.

Leimu R *et al.* 2008. Does it pay to have a "bigwig" as a co-author. *Frontiers in Ecology and the Environment* **6**(8): 410–411.

Levy GN. 1997. Surprise authorship. *Science.* **275** (5308): 1861. DOI: 10.1126/science.275.5308.1861d

Lu Z. 2011. PubMed and beyond: a survey of web tools for searching biomedical literature. *Database*; published online January 18, 2011. DOI: 10.1093/database/baq036

Marder E *et al.* 2010. Impacting our young. *Proceedings of the National Academies of Science* **107** (50): 21233.

Monastersky R. 2011. *The number that is devouring science.* IFORS. ifors.org/web/the-number-thats-devouring-science; published online April 03, 2011.

Rossner M. 2002. Figure manipulation: assessing what is acceptable (editorial). *Journal of Cell Biology.* DOI: 10.1083/jcb.200209084

SCImago. 2007. SJR — SCImago Journal & Country Rank. Retrieved January 05, 2012, from http://www.scimagojr.com.

Singapore Statement on Research Integrity, July 2010, http://www.singaporestatement.org.

Starr J and Gastl A. 2011. CitedBy: A Metadata Schema for DataCite. *D-Lib Magazine* **17** (1/2, January/February).

Steneck, NH. 2010. Perspective: put integrity high on your to-do list. *Science Careers.* DOI: 10.1126/science.caredit.a1000106

Tananbaum G and Holmes L. 2008. The Evolution of Web-Based Peer-Review Systems. *Learned Publishing* **21** (4): 300–306, http://dx.doi.org/10.1087/095315108X356734

Weltzin JF *et al.* 2006. Authorship in ecology: attrition, accountability, and responsibility. *Frontiers in Ecology and the Environment* 4(8): 435–441.

Index